MW00353086

"Dennis Durst's *Eugenics and Protestan[t]*
the history of eugenics by examining re
tional scholarship. Durst explores how
provided a common ground for discour[...]
who embraced eugenic ideals. Based on painstaking research, and grounded in writing that is both nuanced and clear, this book is an important addition to the history of eugenics."

—**Paul A. Lombardo**, Georgia State University College of Law, Atlanta

"The eugenics movement of the early twentieth century has had a profound influence on medical ethics. Though the core ideas of eugenics go back to Aristotle, teachers of health care ethics have rightly traced their recent roots to the late nineteenth century, within social Darwinism and the Industrial Revolution. But where was the American evangelical church in all this? For more than two millennia, Christianity has defended poor, vulnerable, and marginalized individuals, with their value grounded in the *imago Dei*. Christian theology should have pushed back against the worst abuses of negative eugenics that sterilized the 'feeble-minded' and other genetically 'defective' persons. Yet given the interplay of science, theology, and attempts at social reform, eugenics sometimes co-opted the church. Philosopher and theologian Dennis Durst deftly and thoughtfully examines these complex relationships. This book is a rich academic resource for seminary and health care students alike."

—**Dennis M. Sullivan**, Cedarville University, Cedarville, Ohio

"In *Eugenics and Protestant Social Reform*, Dennis Durst documents what happened when bad science and bad theology colluded to create the social monster that was the American and British eugenics movements of the early twentieth century. We who live in the twenty-first century must resist the power and subtlety of emerging biomedical and biotechnological developments that tempt us to embrace a new eugenics. Durst's penetrating analysis can help us avoid repeating the past, if we have ears to hear and eyes to see."

—**C. Ben Mitchell**, Union University, Jackson, Tennessee

"Durst begins by thanking archivists. *Superman* has Lois Lane digging through archives at *The Daily Planet* to uncover a supervillain. Durst documents a history with no mastermind, describing how theology and science 'became entangled' across Protestantism. He shows how eugenics became holy. An 'emphasis on the body' led progressives to regulate 'bodily activities'—to sterilization, institutionalization, incarceration. This book is essential for American Christianity, Ethics, Race Studies, and for anyone trying to live otherwise."

—**Amy Laura Hall**, Duke University, Durham, North Carolina

Eugenics and Protestant Social Reform

Eugenics and
Protestant Social Reform

Hereditary Science and Religion
in America, 1860–1940

Dennis L. Durst

PICKWICK *Publications* · Eugene, Oregon

EUGENICS AND PROTESTANT SOCIAL REFORM
Hereditary Science and Religion in America, 1860–1940

Pickwick Publications
An Imprint of Wipf and Stock Publishers
199 W. 8th Ave., Suite 3
Eugene, OR 97401

www.wipfandstock.com

PAPERBACK ISBN: 978-1-5326-0577-2
HARDCOVER ISBN: 978-1-5326-0579-6
EBOOK ISBN: 978-1-5326-0578-9

Cataloguing-in-Publication data:

Names: Durst, Dennis L.

Title: Eugenics and protestant social reform : hereditary science and religion in america, 1860–1940 / Dennis L. Durst.

Description: Eugene, OR: Pickwick Publications, 2017 | Includes bibliographical references.

Identifiers: ISBN 978-1-5326-0577-2 (paperback) | ISBN 978-1-5326-0579-6 (hardcover) | ISBN 978-1-5326-0578-9 (ebook)

Subjects: LCSH: Eugenics—United States—History | Eugenics—Moral and ethical aspects | Sterilization, Reproductive | Medical ethics | Social Darwinism | Religion and science—History—19th century | Religion and science—History—20th century

Classification: JV6485 d877 2017 (print) | JV6485 (ebook)

Manufactured in the U.S.A.   05/26/17

For Kenric

"Nay, much more those members of the body, which seem to be more feeble, are necessary ..."

—St. Paul, 1 Corinthians 12:22, KJV

Contents

Preface

WITH THE PROFOUND REVISION and questioning among historians of science of a simplistic conflict model of science-religion relations, more complex formulations have become necessary. The relationship between the eugenics movement and religious ideologies, though briefly mentioned at times in the literature, has only recently begun to be explored in depth. Many intellectual streams converged in late-nineteenth-century North America, in a cultural milieu dominated by Protestant theologies, biological theories, and multi-disciplinary efforts at social reform. In the vast river of social mentalities the flow of ideas rarely unfolds along a straight line or in an uncomplicated fashion. By the early twentieth century, the eugenics movement became a flood, inundating the body politic and boasting of enthusiasts across the religious and political spectrum.

Some adherents of Protestant Christianity became promoters and champions of the flow, others began to offer warnings and build flood walls against an ideological tide threatening to swirl out of control. This book is an effort to examine several branches feeding into the eugenic mainstream, exposing degeneration theory as a key tributary. The narrative also strives to highlight some lesser-known efforts to stanch its flow. The interplay of scientific and religious, particularly Protestant, cultural authorities in this story is at times perplexing. Eugenics discourse contained surprising contingencies and odd minglings that seem incompatible to the reader of today. At the outset it is thus important to identify the ideological themes that will recur often within this work.

The narrative centers on heredity, and the shifting definitions and explanations thereof that received free play prior to the time of the rediscovery in 1900 of Gregor Mendel's genetic insights. Holding particular importance were intense debates over the degree of flexibility or mutability of human inheritance. Assumptions about heredity's propensity to be changed through environment and education, or conversely its resistance to such efforts, loomed large in this discussion.

Second, theological anthropology serves a role. Theologians and biblical commentators had much to say about human nature in this era. Internal debates over the definition, extent, and transmission of original sin among Protestants indicated this was a contested domain where resolution required a broader perspective. Theologians often turned to the emerging science of heredity to buttress theological anthropological claims, though they varied greatly over Darwin's thesis of the common ancestry of all living things. Theological engagement with science, even amid internecine theological arguments, was a tacit acknowledgement of the rising authority of modern biology.

As the third and central concern, degeneration theory played a vital role in creating a means by which an older religious and biblical rhetorical tradition began to coalesce with biological reflection. Whereas the Victorian or Gilded Age ethos was often optimistic, positive, forward-looking, as well as theologically postmillennial, this is only a partial and somewhat distorting view. Degeneration theory gave voice to burgeoning fears that many elites harbored about human atavism and the potential for regression. Many of them warned the public about the fragility of modern civilization. Whole families could, by habits or poor breeding, experience reversion and even extinction. Some reformers feared that if such could happen to families, a kind of "race suicide" could befall humanity, having disparate impacts upon the various races according to their alleged fitness for survival. Evidences of degeneration began to proliferate in the thought of scientists, religionists, and social reformers. Cultural shapers began to construct subsets of the population as a cumulative threat to the forward progress of civilization. In order to weed out such threats, experts began to arise who could ostensibly read the "stigmata" or signs of degeneration from the bodies of the feebleminded, criminals, epileptics, alcoholics, or racial minority populations. If such so-called unfit persons could be constrained or prevented from procreation, reformers thought the threat of degeneration could be curtailed, and thus modern society could be saved.

Further, both Protestant churchmen and Protestant scientists lent their energies to social reform efforts partaking of ever more organized and self-consciously scientific and professional levels of efficiency. Many social reforms we today would take as unproblematic and beneficial, such as clean water and child labor reforms. Others appear more prejudicial in retrospect. Family studies, under the auspices of organizations such as the American Eugenics Society and its state affiliates, undertook to examine families for hereditary defectiveness and degeneracy. Professional societies, conferences, and journals proliferated to address the pervasive problems associated with the underclass. The social gospel emphasis on the body more than the soul as the locus of redemption often led reformers to undertake the control of the bodily activities of those who fell short of a vision of perfection. This was especially acute in movements for marriage restriction and eventually the eugenic sterilization of those who appeared congenitally deficient in a variety of ways.

This narrative would be incomplete without acknowledging those who began to oppose the ideology of eugenics. As theories of heredity became more exacting, and as cultural anthropologists studied after a scientific fashion such factors as nurture and education, eugenics began to lose its luster. Eugenicists themselves became a target of ridicule for those who saw them as busybodies or utopian idealists. Further, some Protestants emphasized anew older theological themes that implicitly challenged elitism and the separation of the "unfit" from the "fit" in society. The right to marry and procreate according to personal choice won out over eugenic meddling. Waning support for eugenics by scientists also eroded any automatic deference to its putative scientific credibility among non-scientists.

While notions of the unfit are far more subtly expressed today, they are not completely absent from the twenty-first century. Special efforts to promote birth control among minority populations, or assumptions that the disabled may arrive in the world via "wrongful birth," provide developments for continuing discussion and debate. Theological considerations of a traditional kind can seek to defend the solidarity of all human beings as in some fashion unfit, imperfect, flawed, and in need of love, community, and empathy. Disability Studies is emerging to serve the role of advocacy like never before. The mapping of the human genome indicates that all humans carry within them the potential for both health and disability or disease. Environment indeed plays a role in many of the ways genetic factors come to manifest themselves. Compassion for those suffering at the social margins calls us as responsible citizens and as persons to reflect on the resources of the past as well as the present. History never merely repeats itself. It does however offer insights as well as cautions for how we blend the excitement of our new discoveries with the perspectives gained from ancient sources of wisdom and moral awareness.

Acknowledgments

MANY INDIVIDUALS AND INSTITUTIONS, who often labor quietly and unseen to public view, contribute to the writing life of an academic. It is hard to imagine the craft of writing non-fiction could survive if these unsung heroes in the garden of academe should suspend their diligent attention to the wellsprings of knowledge. I first wish to thank the library staff at Kentucky Christian University, especially Elizabeth Kouns, Naulayne Enders, and Tom Scott. Their patient ministrations and careful watch over my many interlibrary loan requests are worthy of high accolades. Given the disturbing and wide-ranging nature of the titles of said requests, I do wonder what they think this professor could possibly be writing about. Their good humor is surely one of the attributes that makes them stellar colleagues and dear friends.

Many students have assisted me with this labor in between the myriad other duties of overseeing my record-keeping and grading regimen. Katie Longacre spent many hours honing the bibliography, and other workers contributed help as well, most notably Jessica Proudfoot, Staci Crawford, Sarah McVaney, and Sarah Demas.

Several colleagues have read drafts of essays over the years during the period we sustained one another through the KCU Writers' Colloquy. Charlie Starr, Nathan Coleman, Ralph Hawkins, Scott Caulley, and Rob O'Lynn willingly commented on decontextualized chunks of material without even a slight complaint. Nathan especially was instrumental in helping me get my ideas out into the public via *Nomocracy in Politics* online. There the shepherding of editors Peter Haworth and Joshua Bowman enabled me to put my ideas forth in a venue dedicated to discussions of social policy and its historical underpinnings. Their patience with my wide-ranging and quirky interests is a testament to their support of creativity even in the field of non-fiction historiography. My thanks to *Nomocracy in Politics* for permission to include materials previously published by them hereafter, especially in chapters 3, 4, 6, and 10.

Archivists are also essential guides to researchers trying to locate long-neglected primary and secondary sources. Here I owe thanks to many, whose names I never knew or have forgotten, who patiently answered my questions and brought forth their treasures. Archivists at Pius XII Library at Saint Louis University gave me my start on this project. The highly professional staff at The American Philosophical Library in Philadelphia, Pennsylvania assisted my investigation of the American Eugenics Society archival sources, including Eugenic Sermon Contest documents. The United Methodist Archives and History Center at Drew University in Madison, New Jersey, helped me grasp the socio-theological context of Protestant home missions during the Progressive Era. The Hekman Library at Calvin College in Grand Rapids provided a quiet venue for writing and reflection during my continuing education venture in a science and religion Summer Seminar in Christian Scholarship directed by eminent church historian Joel Carpenter. The Concordia Seminary Library in St. Louis was instrumental in helping me navigate the writings of Walter A. Maier. Material on temperance figures was shared by the Willard Memorial Library Archives in Evanston, Illinois.

The Indiana State Archives Commission on Public Records is a treasure trove of eugenics primary materials. I acknowledge this archive for granting access to the Amos W. Butler unpublished correspondence. This material is vital to investigation of his years of service in the National Conference of Charities and Correction. These and other materials in that archive served as source materials for part of chapter 5.

The journal *The Cresset: A Review of Literature, the Arts, and Public Affairs* graciously granted permission to use material previously published on Lutheran theologian Walter A. Maier. Some of this material is found in chapter 10.

Zana Sueme at the Drinko Library at Marshall University in Huntington, West Virginia, was ever helpful in finding titles quickly and efficiently. The Special Collections of the Princeton Seminary Library at Princeton University helped me in my investigations of missiologist Robert E. Speer. The Robert Woodruff Library at Emory University enabled me to gain access to cogent secondary literature in the field of Disability Studies. Numerous public librarians contributed to my research efforts, including St. Louis Public Library with its fine collection of the magazine *Good Health* under the long editorship of John Harvey Kellogg. My local public librarians in Greenup County, Kentucky have always offered help when asked. Such library and archival professionals were always courteous and answered even recondite questions with unfailing patience and grace. Thanks to Aurelia Dyson for terrific last-minute editing advice.

C. Ben Mitchell at the journal *Ethics and Medicine* has been very supportive of research on the history of eugenics and its intersections with religious thought. I wish to acknowledge Ben and the journal for giving a place for discussion of the connections between the rise of eugenics and the development of the field of bioethics. Eminent scholars Paul Lombardo and Ronald Numbers also offered helpful and trenchant criticisms on earlier drafts of this work.

I wish to thank the Louisville Institute for granting me their 2002 Dissertation Fellowship. The Louisville Institute's support of scholarship at the intersection of pastoral formation and academic theology is a unique and vital contribution to the life of both church and university. My doctoral work would have suffered greatly if not for their generous assistance.

My Doktorvater Jim Fisher has always been an encouragement, from the very first day I met him in his office in Verhagen Hall at SLU. He has a huge part to play in this project, and his unstinting encouragement has been vital to my perseverance over many years. I would be remiss if I did not also pay tribute to Charlie and his amazing passion for life and adventure.

As is customary, I absolve everyone but myself for any infelicities or ill-considered conclusions drawn in this book. While every author stands on the shoulders of giants, it is still up to the individual to stand firmly upon the convictions drawn from reading, fellowship, prayer, and life experiences both pleasant and painful. The final and deepest acknowledgement must go to the one who is the author of life itself, life expressed in all of the variegated beauty of the human family. *Soli Deo gloria.*

Introduction

What is man that thou art mindful of him? And the son of man that thou visitest him? For thou hast made him a little lower than the angels, and hast crowned him with glory and honour.

—Psalm 8:4–5 KJV

"Behold, I was shapen in iniquity; and in sin did my mother conceive me."

—Psalm 51:5 KJV

Introduction

SEPARATED BY ONLY A few pages in the King James Version of the Bible we find statements indicating both the dignity and the degeneracy of the human condition. Encapsulated in them is the tension that stands at the heart of Christian theologizing about human nature. Not merely human origins, but human procreation and human heredity became fields for fierce debate in the Christian theological tradition. With the events leading up to the birth of genetics, and efforts to apply genetic research to complex human problems, the interface of theology and heredity took on a new cast. The rise of the eugenics movement, including both its ideological and public policy faces, added to the complexity of answering a fundamental question: "What does it mean to be human?"

Dignity and degeneration are religious and philosophical terms. But the valences of the latter term, degeneration, were altered by the rise of biological and genetic theorizing about how heredity, especially human heredity, functions. This alteration, far from completely abandoning

1

earlier theological discourses on anthropology, co-opted and in some cases distorted theological meanings. This book tells the story of the changing discourses of the theologies and the sciences concerning human heredity during the period of 1857–1939. 1857 represents the year degeneration theory was first promoted in book form, 1939 represents the effective end of the American Eugenics Society and the start of World War Two. Degeneration, a core concept of the eugenics movement, served as a key conceptual nexus between theological and scientific reflection on heredity among Protestant intellectuals and social reformers in the late nineteenth and early twentieth centuries.

The Core Questions

It is a commonplace maxim that the nineteenth century was a century of great change, and that old verities were cast aside and new ideas supplanted them in radical ways. But this template ignores the importance of assessing both continuity and change over time in the post-Darwinian West. Especially when assessing changes in theology, within which theologians have a substantial vested interest in the preservation, transmission, and re-appropriation of biblical and confessional ideas, it is best to take the long view. Questions raised by the commingling of theological and scientific notions of human heredity and its defects include the following.

What were the major terms by which scientists described human inheritance? Was heredity strongly fixed and resistant to modification, or was it malleable relative to environmental influences? How did fears about the fragility of the human "germ plasm" become amplified in this period? What role did atavism or reversion play in the imaginations of elite thinkers reflecting on social problems like mental disorders or criminality?

How was religious rhetoric or theological reflection, including allusion to the Bible, applied to specific social problems in this era? How were those families eugenicists deemed problematic described in family studies literature? How were persons with mental illness or mental challenge described? What rhetoric was employed to amplify social fears about such persons? What solutions were attempted in order to rein in the putative problem of hereditary mental defect?

How were criminals viewed by social reformers who embraced the ideology of degeneration theory? How was biblical or religious rhetoric deployed in promoting the notion of hereditary criminality?

What role did assumptions about heredity play in issues such as cognitive disability, immigration restriction, race relations, alcohol abuse, and marriage policy in the USA? What assumptions drove elites to assert social control over these areas of public life, and what forms did this assertion take? What role did involuntary sterilization come to play in social policy proposals forged to address such issues?

Finally, what legacy did the blending of religious and hereditary themes bequeath to the twenty-first century? Are there lingering streams of eugenic ideology still at play in the contemporary world?

Degeneration Theory and Eugenics

Degeneration theory may be traced to the French psychiatrist Bénédict Morel (1809–1873) in the mid-nineteenth century. Originally embedded in theological categories of creation and the fall, scientific investigations of generation and degeneration have never been free of the presuppositions that ground research agendas across the disciplines. Degeneration theory had obvious strong theological and biblical antecedents, not least of which was a reinterpretation of the Christian doctrine of original sin. Degeneration language invited a strongly moralistic assessment of humans, especially those situated on the margins of society. According to one student of Morel's thought, "Morel founded his concept of degeneration on a theological conception of creation and on an ethical conception of the moral law." For Morel, human pathology and normality were both equally defined based upon "the philosophical and theological problematic of the relation between the soul and the body."[1] In a trope that would deeply trouble future eugenicists, Morel believed that if two individuals sufficiently degenerative in physiology mated, their descendants would deteriorate into a subhuman status and eventually become sterile.[2]

In a study of the interplay of theology and science at the nineteenth-century intersection of medicine and morality, Rimke and Hunt note that reformers such as British psychiatrist Henry Maudsley (1835–1918) applied Morel's theory in an explicitly theological way:

> A central role was played in degeneracy theory by the theologically laden dictum of "the sins of the father"; the significance being that all were vulnerable, potential bearers or receivers of degeneracy: "each individual, each family, each nation may take either an upward course of evolution or a downward course of degeneration."[3]

1. Liégeois, "La Théologie," 356, my translation.
2. Morel, *Traité des Dégénérescences,* 683.
3. Rimke and Hunt, "From Sinners to Degenerates," 59–88.

Morel did advance the field of psychiatry by identifying physiological causes for certain brain disorders. He also stressed a balance of environmental and hereditary factors in contributing to mental illness. Others would later embrace degeneration in more ideological terms, and move away from Morel's balanced nature plus nurture approach. Eugenics-minded social reformers, while deploying both scientific and religious resonances of the term "degeneration," would come to place greater weight upon heredity, often to the neglect of the influence of environment on social problems.[4]

The major contours of Daniel Pick's analysis of degeneration in the European context readily applies to the American setting as well. Pick notes that documentation of diagnoses in French and British mental hospitals of the latter nineteenth century routinely began with mental degeneration, followed by one of a host of more specific pathologies. The widespread usage of degeneration theory occurred not merely in biology or psychiatry, but in various genres of social commentary such as journalism, drama, and political discussions. This contributed to a widespread cultural dread of degeneration as "threatening the very overthrow of civilization and progress."[5] Pick explores degeneration theory in the United Kingdom, but little has been written on degeneration to include the North American religious context. For instance, one major historical treatise whose editors include scholarship from various experts on degeneration theory covers such diverse topics as anthropology, sociology, psychoanalysis, biology, medicine, technology, political theory, literature, art, and theater. No chapter, however, is devoted to the connections between degeneration theory and religion.[6]

While various dimensions of the American eugenics movement have been surveyed and debated in the secondary literature, relatively little attention has been paid to the theological or religious interactions with the emerging scientific discourses on heredity.[7] The existing efforts to address the historical data tend toward the simplistic. So in an otherwise fine overview of the sterilization movement in early twentieth century America, historian Barry Bruinius writes: "Similar to the old Calvinist and Puritan ideas of the chosen and the damned, eugenics thinkers, wielding the banner of evolution, divided the world into the manifestly unfit versus fitter families, degenerate offspring versus better babies, and citizens of the wrong type

---

4. For a succinct overview of degeneration theory in the field of psychology as well as in popular culture in the late nineteenth century, see Gamwell and Tomes, *Madness in America*, 124–31.

5. Pick, *Faces of Degeneration*, 9.

6. Chamberlin and Gilman, *Degeneration*, 1–303.

7. See Kevles, *In the Name of Eugenics*, 61, 68, 118; and Rosen, *Preaching Eugenics*, 3–24.

versus the hearty American stock."[8] The problem with this analysis is that eugenics departed from central tenets of the teachings of Calvinist and Puritan thought in a radical way. For example, the Calvinist and Puritan ideas of the chosen and the damned refers to the doctrine of election in the Reformed tradition. Bruinius assumes substantial continuity between the fit/unfit dichotomy in eugenics and the saved/damned dichotomy in Puritan theology. Yet any careful reading of either Calvin or the Puritans would clarify that election was emphatically not based upon the healthiness, goodness, or well-born (eugenic) status of the elect. Theological election is, under this interpretation, based in the inscrutable decree of the Almighty. One need not be overly sympathetic with this theological tradition and its way of parsing out the doctrine of election to recognize that its connection with eugenics is tenuous. Both the dissimilarity and the similarity between eugenics and the Calvinist tradition are worthy of investigation. This theological change over time, therefore, is vital to our understanding of the use and abuse of those theological themes we discover in the eugenics literature. This dissimilarity illustrates how theological distortions, not merely theological discourse *per se*, contributed to the human rights abuses we find troubling today.

The literature on Darwinism and religion, including the Protestant intellectual response to the theory of evolution, is vast. Historians Ronald L. Numbers, John Hedley Brooke, Jon Roberts, Edward J. Larson and James Moore top the reading list of the student wishing to explore Darwinism and its religious reception in the transatlantic milieu. Perhaps because it did not occasion the same fervent controversy at the time, religious responses to eugenics have received much less attention in the secondary literature. Though the founder of eugenics, Francis Galton, was a cousin of Charles Darwin, the line from natural selection theory to the eugenics movement is a crooked line at best, the historical telling of which has not often been sufficiently nuanced. The topic of this book, eugenics and religion, first suggested by Galton himself, has garnered somewhat rare attention. The most important book in this field is Christine Rosen's *Preaching Eugenics* which is an excellent overview of churchmen and their support of the eugenics movement. My project goes deeper into the nineteenth-century theological and ideological antecedents of the mixture of religion and eugenics by attending to degeneration theory as an important foundational trope in the rise of eugenic social policies. I seek to delve into the thought of a wider range of social reformers including, but not limited to, clergy.

---

8. Bruinius, *Better for all the World*, 239.

The eugenics movement arose as the most evident social reform impulse spawned by the proliferation and diffusion of degeneration theory and rhetoric of degeneration. More specifically, degeneration theory underpinned the concept known as *negative eugenics* or "the belief that some people were physically, mentally, or socially degenerate by virtue of defective biology and that those dysgenic traits, passed on to offspring, would lead to racial degeneracy (weakening of the human race)."[9] Many Protestant progressives enthusiastically supported and participated in the eugenics movement. The discourse of degeneration as sin (theology) and degeneration as biological regression (science of heredity) became entangled, leaving many citizens in its wake having been tendentiously identified as "mental defectives," "hereditary criminals," and "moral degenerates." For some this stigma led to their involuntary sterilization; for others the denial of the right to marry; for still others it meant prolonged unwarranted stays in mental institutions or prisons. Reflective of the broad reach of eugenics, the sources for this book are quite diverse. Here I explore systematic theology textbooks, Protestant journals, marriage manuals, published sermons, public policy documents, charities and corrections conference proceedings, race betterment conference proceedings, eugenics promotional books and articles, social reform and penology literature, temperance literature, and scattered remarks through the primary periodical literature discussing science and religion. While the theme of degeneration is found mentioned throughout eugenics primary literature, its theological and religious connections shall be the driving focus in this book.

Renegotiations of the Cultural Authority of Ministers and Theologians

The Christian tradition is well-known for promoting two fundamental doctrines relevant to this project: creation and fall. The post-Darwinian controversies over the first term, creation, have received exhaustive treatments elsewhere, and shall be touched on only briefly in the following pages. By contrast, the fall and the concomitant though controversial doctrine of original sin has not been adequately explored in the post-Darwinian literature. There are some good recent overviews of the doctrine of original sin as a cultural meme, but not too much attention has been given to its cross-disciplinary science-theology resonance *at the time prior to and including the birth of genetics as a field.* This study aims to stimulate a greater discussion of this connection.

9. Engs, "Negative Eugenics," 160–61.

During the period of the 1860s through the 1930s, religious and scientific discourses blended and clashed in unique ways to forge ideologies of heredity and the social policies built upon them. This book seeks a window upon the ways scientific debates over heredity influenced, or were influenced by, the theological and social reform literature of Anglo-American Protestant elites. While the European continental background is very important, and will come into play, the American context is my primary focus. Such a book could never be an exhaustive treatment of this vast subject matter, but serves rather as a series of topical soundings intended to foster deeper research agendas and further many conversations across various disciplines curious about history, science, religion, theology, and social policy.

The historiography of Protestantism shows that traditionalist and progressive wings of a broad Protestant mainstream conducted an often vituperative form of discourse. Routinely, theological storms occurred over the findings and methods of science, as well as philosophical disputes about the role of science in shaping religious assumptions. The fundamentalist-modernist controversy that shook many Protestant denominations continues to complicate historians' efforts to assess the relationship between scientific and theological sources in Protestant intellectual life.[10] While it would be tempting to believe that conservatives rejected eugenics due to skepticism toward its scientific basis, this was not universally true. While it would also be tempting to believe that all progressives embraced eugenics because it seemed (at least before the late 1920s) to have the *imprimatur* of science, this also was not universally true. Evolution was a fairly clear dividing line between conservatives and progressives in Protestant thought; eugenics, comparatively speaking, was not. Why eugenics could gain adherents across the ideological spectrum may be answered, in part, by paying attention to the theological sub-stratum undergirding investigations of heredity in this period.

The literature on the relationship between science and religion in the post-Darwinian era is vast. Thus a point of focus is essential to progress to explore the relationship as it relates to the rise of the "religion of eugenics" desired by Francis Galton, Charles Davenport, and other prominent eugenicists. The most useful lens through which to observe the science/religion interface *vis a vis* eugenics is the lens of *cultural authority*. This lens was comprised of the shift of cultural authority from the traditional religious moorings of mainstream Protestantism to the rising natural sciences. What is vital here is to understand this shift not in terms of a hostile takeover of cultural authority by scientists in an epic battle with recalcitrant obscurantist

10 See Szasz, *Divided Mind*, 92–125.

defenders of religion. That would be to reinscribe a now-discredited warfare model of the science/religion relationship. More accurately, we must understand this shift in terms of the willing acquiescence, and indeed enthusiastic embrace of new findings of science by major figures and gatekeepers of the religious establishment. An exploration of this phenomenon will occur in chapter 10 of this book.

As the gradual shift of cultural authority from theology to science occurred, and as clergy scrambled to accommodate science into their approach to social problems, a valuable critical distance was sacrificed. This loss of cultural authority then opened the door for eugenics and the human rights abuses that came to characterize social policy. The traditional role of the clergy had been to stand apart from society with some attempt at detachment, to offer objective critique, or at least an occasional vocal and prophetic resistance. Apart from Roman Catholic objections, accommodation became the default position of many theologians and ministers toward science, and the abuses of human rights inculcated by eugenicists were able to develop with little resistance.

A Brief Overview

The thematic breadth of the reach of eugenics may be discerned by a brief survey of the chapter topics developed here. Chapter 1 explores the connections between Bénédict Morel's degeneration theory and the rise of eugenics. The potency of this concept for bridging between older religious discourse and the new discourse of biology is illustrated and explored. The popularity of literature on atavism as a form of degeneration set the stage for public fears over categories of persons seen as hereditarily defective.

Chapter 2 looks at the links between theories of heredity and the rise of the eugenics movement. Figures such as Francis Galton and Charles Davenport promoted eugenics as a new religion or new creed. The debates between followers of Lamarck and followers of Weismann indicated that the science of heredity had not settled the nature/nurture debate. Herbert Conn, a biologist at Wesleyan University promoted the notion of cultural evolution to counter-balance the increasingly hard view of heredity and the biologically deterministic emphases of the eugenics movement. Yet insights into culture and the impact of environment on complex social problems like those articulated by Conn could not stop the development of social policies based on eugenic assumptions.

Chapter 3 briefly explores the popular notion that mental and even moral defect can be read from bodily features of allegedly "unfit" citizens.

The religiously-laden term, "the stigmata of degeneration," took on a multi-layered cluster of meanings through the writings of criminologist Arthur Macdonald. Such a subversion of the term *stigmata* morphed into the outright stigmatization of the poor, of the cognitively disabled, and of those identified as "hereditary criminals" during the heyday of eugenics.

Chapter 4 elucidates the notion that elites could detect defectiveness and degeneration by scientific investigation of families. In this period the rise of the eugenic family studies literature indicates that not merely individuals but entire family lineages became suspect in the eyes of social reformers. Thus 'nests of incompetence' could be identified, categorized, and brought increasingly under social control through institutionalization and even sterilization.

Chapter 5 focuses on the theme of the "menace of the feeble-minded" in the early decades of the twentieth century. This trope arose as notions of hereditary and multi-generational mental defect, aided by intelligence testing and eugenic assumptions about morons, idiots, and imbeciles took hold of the public imagination. The social policy impact of such prejudicial rhetoric toward the cognitively disabled took the form of segregation of the sexes in mental institutions, and involuntary sterilization of those deemed unfit to procreate.

Chapter 6 is an exploration of one category of persons erroneously identified as mentally defective and their treatment by social reformers. Persons afflicted with epilepsy were routinely lumped in together with persons with severe mental incapacitation, despite the oft-acknowledged fact that epilepsy could be found among famous persons of great genius. This chapter thus illustrates ways that epilepsy and its treatment rendered problematic the category of congenital mental defectiveness and its role in a totalizing eugenics discourse.

Chapter 7 explores the rise of eugenic criminology and the waxing and waning of the concept of the "born criminal," as both a sociological and a theological problem. The prominence of clerically trained criminologists, penologists, and sociologists indicate the religious roots of eugenic assumptions about criminals. The role of nurture and environment in the etiology of criminality began to gain prominence as hereditarian notions of the born criminal became increasingly scientifically suspect.

Chapter 8 indicates that the interplay between the temperance movement and the eugenics movement was no accident. The rise of "scientific temperance" indicated that cultural authority for the promotion of prohibition was not uniquely driven by religious considerations. Rather, temperance reformers, moved by religious conviction, sought out scientific evidence to buttress the case against alcohol use. Concerns about the impact of alcohol on the human "germ plasm" served as a key connection point between temperance advocacy and eugenics. Several reformers in the Woman's Christian Temperance Union wrote about procreation. Such literature highlights the interface of science, religion, eugenics, and temperance.

Chapter 9 connects eugenics and religion with the history of scientific racism. Notions of inferior and superior races, the ideology of Nordic supremacy, and the movement to restrict immigration by those of allegedly "inferior stock," all drew upon eugenic themes. By restricting immigration, elites hoped they could avert "race suicide," and prevent the "passing of the great race." Some theologians, such as Augustus H. Strong, reinforced racial biases using degeneration theory as a resource. But based on convictions derived from the Bible, theologians such as John Miley, James Boddy, and Robert Speer insisted on the unity of the human race, and resisted notions of racial superiority for whites.

Chapter 10 explores further the ways several theologians reflected on the rising theories of biological inheritance, and sought to employ them in discussions of ancient themes such as sin and its transmission. One theologian, Walter A. Meier, wrote extensively about eugenics, and urged caution against its more enthusiastic iterations in public policy discussions in the twenties and thirties.

My conclusion suggests further lines of research that may portend troubling continuing eugenic attitudes in the medical and social policy fields. Debates over wrongful birth lawsuits, the selective aborting of unborn children with Down Syndrome and other disabilities, and procreation restriction efforts in minority communities, are worthy of consideration and study in light of the cautionary tale of the rise and fall of eugenics. Theological awareness of the solidarity of the human race in a shared transcendent dignity, amid an acknowledgement of the universal experience of genetic imperfection, can lead to a greater humility and acceptance of differences. The intersection of theology and disability studies is an encouraging development explored briefly, and gives hope of a more sensitive and inclusive vision of society.

Rhetoric and Reform

The contested domain of human frailty, human disability, human weakness, and human depravity took on first a biological and then a sociological cast at the turn of the twentieth century. While religious folk were often optimists about the forward march of civilization, there was always the foreboding counterforce of a conviction that the human race had a natural propensity toward evil, and toward reversion to a primal, uncivilized state. Theologians and scientists were people of their time, and only a few were satisfied to dwell exclusively in an ivory tower. They often interacted with social reform and policy-making, and even became reformers themselves, to varying degrees. The rhetorical and intellectual resources of religious, theological, and biblical language were readily deployed by scientists who were Protestant adherents of varying denominations. So diffuse was this linguistic toolbox that even elites with little or merely vestigial religious faith still could utilize them as part of an overall argument trajectory.

One lesson to be gained through this project is to urge great caution when either theological or scientific ideologies are brought into social policy debates, and especially when any convenient marriage thereof makes an appearance. Like learning of a sibling's Las Vegas wedding, both families might rightly be skeptical of any such "match made in heaven." Certainly the match, especially for those involuntarily sterilized, or institutionalized on dubious scientific grounds, or cruelly denied a marriage license, seemed to have been made in the deep recesses of another place.

CHAPTER 1

Degeneration Theory and Eugenics Discourse

These facts show that heredity is not a freakish or haphazard principle but a great and immutable law, as positive and certain in its action as the law of gravitation. The lesson which we desire to draw from the knowledge of this great biologic principle is the fact that in all civilized lands heredity is taking note of every condition, every habit, every act which results in the deterioration of the stamina of the race. Nature is a good bookkeeper; nothing escapes her notice.

—John Harvey Kellogg, 1910[1]

Introduction

FRENCH PSYCHIATRIST BÉNÉDICT MOREL (1809–1873) was a pioneer in integrating biological explanation of human defectiveness with psychiatry. Though he was a devout Roman Catholic who acknowledged the role of spiritual factors in mental health, Morel also stressed the biological causation of many forms of insanity. For Morel, the disease model formed the best framework for explaining mental disorders. As a monogenist (literally: one beginning), Morel accepted the created unity of the human race, and in his 1857 *Treatise of Degeneracy* set forth degeneration as a "morbid deviation from the primitive human type."[2] Morel accepted the role of original sin in his notion of degeneration, but saw this as a universal problem and thus not a sufficient explanation for the level of degeneration seen in his patients. According to Coffin, original sin in Morel's view "needs to be associated with

1. Kellogg, "Mendel's Law," 736.
2. Coffin, "Heredity," 156.

other causes coming from the pathogenic natural and social environment to carry on influence."[3] Morel pioneered an investigation of the role of brain lesions within the development of insanity. He blended milieu (his term for environment) with hereditary factors in explaining the multi-generational effects, for example, of alcoholism. Coffin notes Morel's observation that "Individuals drink because of their moral misery and drinking has negative physiological impacts on the body. Then, children from this heritage are particularly exposed to degeneration. Organic predisposition and heritage of a vicious milieu will indeed favor the disease."[4]

Degeneration Theory as Nexus between Science and Religion

The interplay of science and theology is a marked characteristic of Morel's theory of degeneration. Coffin summarizes it succinctly: "Milieu and the original sin cause abnormality; heredity maintains and accelerates the abnormality and there is no cure because this is the tragedy of the human race."[5] Many later social reformers would use the language of degeneration while moving away from Morel's resignation in the face of this tragedy. They would seek out eugenic cures not primarily for the benefit of individual patients but for the human race (or the "germ plasm") as a whole.

Not far behind any scientific or theological discussion of heredity and the details of generation comes the tale of degeneration.[6] Like many terms of its day, degeneration came to be used in a wide range of intellectual discourses, with a resulting swelter of differing and contradictory meanings. At the center of the term's meaning, degeneration stood out as "a condition analogous to illness in which the human organism is said to exist in a state of decay." Further, "the deterioration can be physical, intellectual, psychological, or all three at once."[7]

For religious figures the term usually carried connotations of sin or moral deficiency. Degeneration theorists looked for various signs of degeneration, such as cranial or facial characteristics. This propensity was in part due to the contemporaneous popularity of phrenology, the study of cranial shapes to discern mental ability. Those who imbibed degeneration theory

3. Ibid., 157.

4. Ibid., 159.

5. Ibid., 161.

6. For Morel's influence, see Coffin, "Le Théme de la Degénerescence," 727–29; and Rosenberg, *No Other Gods,* 43–53. See also Gilman, "Degeneracy and Race," 30–35.

7. Gibson and Rafter's definition in: Lombroso, *Criminal Man,* 404.

also held a wide range of maladies to be evidence of degeneration, such as epilepsy, criminality, pauperism, alcoholism, and a host of other social ills. Such alleged evidences of degeneration were often designated by the religiously-laden term "stigmata" of degeneration, the extensive subject of a later chapter.[8]

Degeneration theory echoed theological and biblical antecedents, not least of which was the Christian doctrine of original sin. Degeneration theory also invited a strongly moralistic assessment of humans, especially (for most social reformers) "other" humans, mainly of the underclass. Degeneration theory also got caught up within racialist theorizing among scientists and theologians alike, to the disfavor of minorities and immigrants.[9]

The widespread usage of degeneration theory, not merely in psychiatry, but in various genres of social commentary such as journalism, drama, and political discussions, led to a widespread sense of degeneration as "threatening the very overthrow of civilization and progress."[10] Pick notes that degeneration is not a new concept, tracing it back at least as far as the pre-Christian figures like the satirist Horace or the philosopher Plato. He insists that the particular historical form the theme of degeneration took in the nineteenth century is worth investigating, noting that "important and revealing differences do emerge in the meaning of seemingly homogenous and timeless concepts."[11] Pick's analysis of Morel's studies of cretinism in the 1850s illustrates how Morel came to espouse a degeneration theory wide in social application. Pick writes: "The cretin was an instance, an emblem of racial degeneration. The notion of degeneration became more elastic and expansive, whilst the prognosis of the degenerate became more rigid."[12] Transatlantic fascination with heredity would only grow in the ensuing decades, and versions of degeneration theory spread across many disciplines. The discordant ways degeneration was deployed by Protestant theologians and scientists in the American context add yet more layers to a multi-layered historical narrative of degeneration.

The religious resonance of the pejorative noun "degenerate" has long been seen as a sermonic synonym for an especially depraved sinner. The religious identification of the person's very being with their most obvious flaw, along with the quasi-scientific resonance of degeneration as biological

---

8. Talbot, *Degeneracy,* 33.

9. As Stuart Gilman has argued, degeneration theory was also utilized to lend scientific legitimization to the notion of inherited distinctions between the races. Gilman, "Degeneracy and Race," 27.

10. Pick, *Faces of Degeneration,* 9.

11. Ibid., 18.

12. Ibid., 48.

defect or regression, makes the sources ripe for conceptual slippage and distortion. The proliferation of degeneracy language, the development of its use in the discourses of theology and science, and its importation into social policy, served a recurring role in the history of all three topics. Exploration of degeneracy, with its multivalent nuances and usages in historical context, can shed light on how distortions of both theology and science can easily occur when poorly conceived terms with rhetorical power come to serve as the basis of social policy.

Degeneration and Atavism

In an era dominated by notions of progress, the spectre of atavism, or finding "throwbacks" to an earlier, retrograde, semi-human, and eventually bestial form, haunted the public consciousness. Chicago Physician Eugene Talbot asserted in 1898 that "Reversional heredity or atavism consists in the reproduction in the descendants of the moral or physical qualities of their ancestors."[13] Talbot was one of Morel's leading American interpreters, and embraced a definition of degeneration as "a marked departure from the original type tending more or less rapidly to the extinction of it." Talbot was skeptical of Morel's religious starting point, namely that "there was a primevally perfect man" (i.e., biblical Adam). He quipped that, "while Morel practically outlined the modern study of degeneracy, his theologic timidity forced an absolute definition of a state which, according to his own admission, was purely relative."[14] While disdainful of the theological nuances of degeneration, Talbot would nonetheless interpret theology's role in the influence of degeneration theory. "Precedent to 1835 determinism in popular thought due to Calvinistic predestination had, in English-speaking countries, fought for the doctrine (of degeneration)"; Talbot conceded. He continued: "subsequent thereto the theologic reaction against Calvinism was a strong opposing force, whose influence was finally destroyed" by the triumph of evolution in the 1870s.[15] Thus Talbot was gratified that Calvinism had ostensibly unwittingly promoted degeneration theory, but was chagrined by Calvinists' general opposition to Darwinism.

Leaving aside Talbot's facile blending of theological with biological theories of predestination, the central point remains, namely, that late in the nineteenth century, heredity and theology could find a common discursive frame of degeneration theory within the thinking of American Protestant

13. Talbot, *Degeneracy*, 46.
14. Ibid., 9.
15. Ibid., 26.

elites. Physiological evidences of decay evinced for them an underlying degeneracy that was both biological and moral. Such blending of biological with theological evaluations occurred repeatedly in late nineteenth and early twentieth-century texts, with tragic ramifications for those deemed "unfit" by eugenicists.

Just prior to the rediscovery Gregor Mendel's fundamental principles of genetics and a verifiable set of hereditary laws of biology, the easy conjunction of moral with physical qualities made for a ready blending of theological and biological discourses. Social reformers, especially those enthused by the social control of procreation offered under the aegis of eugenics, would only add impetus to this trend of mixing of morality and physiology. As Thomas C. Leonard has noted, "the new discourses of eugenics and race science recast spiritual or moral failure as biological inferiority, making old prejudices newly respectable and lending scientific luster to the arguments of critics and defenders of American economic life."[16]

One case of the blurring of moral and biological categories was the notion of atavism. Literary historian Dana Seitler notes that atavism "names a condition of 'resemblance to grandparents or more remote ancestors rather than to parents.'" She further points out the historical importance the term, namely: "Atavism brings the ancestral past into conjunction with the modern present, and given the post-Darwinian moment in which the widespread deployment of the concept occurs, this ancestral past was always understood as part and parcel of the course of evolution."[17]

The popularity of novels such as *The Strange Case of Dr. Jekyll and Mr. Hyde* by Robert Lewis Stephenson, *The Monster* by Stephen Crane, the popular series on the atavistic character *Tarzan* by Edgar Rice Burroughs, all the way to the degenerative creature Gollum in J. R. R. Tolkien's *Lord of the Rings* saga, illustrate varied iterations of this foreboding.[18] Degeneration theory gave such inchoate fears anchorage in science and its cultural authority. Outside the realm of fiction, degeneration was taken so seriously it was part of the ostensible justification of various stringent social policy measures to rein in supposedly atavistic or degenerate individuals or families.

While atavism could have the connotation of a kind of abrupt and dramatic throwback to a troglodytic or lower primate state, degeneration was often seen as more gradual, and afflicting broader categories or groupings of people. Degeneration was a diffuse and insidious threat that affected generations of the human race over time. The employment of degenerative

16. Leonard, *Illiberal Reformers*, 124.

17. Seitler, *Atavistic Tendencies*, 2.

18. Edwards, *Gothic Passages*, 81–87; and Seitler, *Atavistic Tendencies*, 1–30.

tropes enabled social reformers to gain political traction for social policy solutions aiming to constrain the fecundity of those identified as degenerate. The occasional monstrous individual was more dramatic than the average degenerate individual, and thus a favorite literary theme. Atavistic or "monstrous" individuals came to be seen as an anomaly so decisively "other" that amusement more than fear was the standard reaction. Witness the popularity of the vaudevillian "freak show" during the same period; wherein such persons were regarded as objects of laughter or pity more than genuine fear. When large-scale groupings of persons became perceived in the public imagination as subject to degeneration, however, this fear became more profound and diffuse, and institutional leaders formulated strategies and policies of social control.

Mendelism, once it became well known, was enthusiastically enlisted by many social reformers as a scientific argument for eugenics, though distorted by multiple oversimplifications of the scientific arguments of Mendelian genetics, as some contemporary geneticists pointed out. Indeed, the Progressive Era momentum of many eugenic social policies ostensibly rooted in hereditary science, can only be understood in light of pre-Mendelian hereditary concepts such as degeneration theory. Ironically, ideas from mid-nineteenth-century figures such as a French Catholic psychologist (Morel) and an Austrian Catholic monk and scientist (Mendel) were often utilized (and conflated) by twentieth-century Protestant reformers zealous for the progress of civilization. Protestant anti-Catholic polemics even included criticism of the supposed dysgenic effects of the vow of celibacy, which was seen to inhibit a desired increase of the "good stock" that the average priest was purported to represent.

After Morel the theory of human degeneration was increasingly shaped, around the turn of the century, by the emerging sciences in Europe and America dealing with heredity, such as cell theory, physiology, and genetics. Degeneration as a theme was disseminated at the popular level through various fields such as biology, medicine, social science, sociology, criminology and religious education. The challenge of understanding these developments requires an exploration of the scientific debates over heredity that emerged during this epoch. To this subject our narrative now turns.

CHAPTER 2

Theories of Heredity and
the Rise of Eugenics

The great problem of civilization is to secure a relative increase of the valuable as compared with the less valuable or noxious elements in the population. This problem cannot be met unless we give full consideration to the immense influence of heredity.

—Theodore Roosevelt, 1917[1]

Introduction

IN ORDER TO UNDERSTAND how the restrictive policies of eugenics came to fruition, we must reach back into the nineteenth century to recognize various streams in the history of human reflection upon heredity. The Mendelian theory that eventually became the consensus view was not dominant until after other less cogent theories had taken root and had already produced social consequences. With the ascendancy of Mendelian laws of heredity, social reformers who already had a stake in restricting procreation of many groups of people identified as "unfit" used Mendelian science as one more argument for a range of procreation restrictions. In truth, adherents of the same biological theory held quite divergent views on what sorts of public policies ought to be based upon, or justified in terms of, that theory. Teasing apart the varying new theories of heredity and their influences on social policy includes looking at how Protestant social reformers used science to undergird efforts at formulating social policy.

Children inherit traits from their parents and grandparents. Yet why some traits are morphologically evident and others are either not present or

1. Roosevelt, *Foes of our own Household*, 257.

18

too subtle for surface observation leads humans curiosity toward the mysteries of heredity. We may add to this curiosity the heartache of parents at the misbehaviors of wayward children that would become instantiated into the increasing paternalism of the Progressive Era social reforms endeavoring to control human procreation, based on particular readings of the science of heredity of the time.

Galton's Eugenics

Francis Galton coined the term "eugenics" in 1883. Eugenics has been variously defined, but the core concern is producing "good births" (the literal meaning of the Greek etymology of the term) based on criteria assumed to be scientific, but often reflective of socio-cultural biases held by upper class Northern European and North American whites. Positive eugenics meant the encouragement of the procreation of those deemed fit to procreate; negative eugenics meant the discouragement of the procreation of those deemed "unfit."[2] Galton concluded his study by admitting the broad scope of his agenda: "The chief result of these Inquiries has been to elicit the religious significance of the doctrine of evolution." He promoted Darwin's insights not merely as a descriptive thesis of common ancestry, but as a prescriptive exercise of moral duty, "an endeavor to further evolution, especially that of the human race."[3]

Galton was a lifelong student of hereditary genius, making his mark by publishing a book under that title in 1869. Human mental prowess and threats thereto remained a major focus of his writings. In 1908, near the end of his life, Galton revised his 1883 work *Inquiries into the Human Faculty* in an effort to sum up his wisdom on the role of human heredity in human genius and human mental defect. For Galton these insights had nationalistic and racial implications, and the ethnocentrism of his views emerged early in the book:

2. The following definitions illustrate the core concepts of eugenics: Kline, *Building a Better Race*, 13: "The science of improving human stock by giving 'the more suitable race of strains of blood' a better chance of prevailing speedily over the less suitable"; Ludmerer, *Genetics and American Society*, 2: "A science which investigates ways to improve the genetic condition of the human race; a program to promote such improvement"; Black, *War against the Weak*, 18: "The study of all agencies under social control which can improve or impair the racial quality of future generations"; and Kevles, *In the Name of Eugenics*, xiii: "The 'science' of improving human stock by giving 'the more suitable races of blood' a better chance of prevailing speedily over the less suitable."

3. Ibid., 220. The implications of this "duty," when seen in the later context of Nazi Germany, is chilling indeed. See Weikart, *Hitler's Ethic*, 31–54; Black, *War Against the Weak*, 247–371; and Kühl, *Nazi Connection*, 68–70.

The moral and intellectual wealth of a nation largely consists in the multifarious variety of the gifts of the men who compose it, and it would be the very reverse of improvement to make all its members assimilate to a common type. However, in every race of domesticated animals, and especially in the rapidly-changing race of man, there are elements, some ancestral and others the result of degeneration, that are of little or no value, or are positively harmful . . . . we are justified in roundly asserting that the natural characteristics of every human race admit of large improvement in many directions easy to specify.[4]

Here Galton made a distinction between harmful elements that are deeply ancestral and harmful elements that occur as the result of a more proximate degeneration. Degeneration could be linked to the reappearance of more primitive features in future progeny. Galton spent many pages arguing that facial features could be categorized and used to identify underlying defects, including problems ranging from criminality to consumption. He sought out the "central physiognomical type of any race or group" by a method of composite portraiture. Galton believed that facial features were so true to type that a composite sketch could be used as a template for identifying and grouping individuals by the resemblance of select facial features.[5] Galton then applied a similar method to identifying and categorizing body qualities. He expressed dismay that among his countrymen, thus: "there can hardly be a sadder sight than the crowd of delicate English men and women with narrow chests and weak chins, scrofolous, and otherwise gravely affected. . . ." Galton laid out the role of heredity and the possibility of improvement by applying the methods of the animal breeder while lamenting: "our human civilised stock is far more weakly through congenital imperfection than that of any other species of animals, whether wild or domestic."[6]

Like Charles Darwin, Francis Galton emphasized sexual selection as an important feature in improving the human race. Heavily gendered biases became manifest in his phrases describing females as "capricious and coy" and possessing "less straightforwardness than the man." Galton saw competition for female affection among males essential to quality control of the eventual offspring. Absent sexual selection, indiscriminate breeding would take over and the results would be negative for the population over time. "The drama of courtship, with its prolonged strivings and doubtful

4. Galton, *Inquiries into Human Faculty*, 2.

5. Ibid., 10.

6. Ibid., 16.

success, would be cut quite short," he mused, "and the race would degenerate through the absence of that sexual selection for which the protracted preliminaries of love-making give opportunity."[7] Hardly a romantic, Galton saw human procreation and animal husbandry alike through the lens of science. The production of characteristics chosen by eugenicists would require a deeper understanding of the science of heredity, and Galton and his peers made the subject a near obsession.

Neo-Lamarckism and Weismann

By the 1880s two major schools of thought rose to the fore in the investigation of heredity with antecedents stretching back at least to the eighteenth century. *Hard* heredity was the position that the unit characters passed from parent to child were particulate and for the most part fixed and resistant to change. *Soft* heredity, by contrast, theorized that the process of development and gestation heavily influenced unit characteristics of an embryo. This position was often labeled *developmentalism* or *transformism*.[8] The most influential proponent of the developmental model of heredity was J. B. P. A. de Monet, Chevalier de Lamarck. Lamarckism has become shorthand for the theory of "the inheritance of acquired characteristics." Under this model, an organism undergoes a change in morphology due to the challenges of its environment. Such changes then become part of the character set passed on, at least partially, to the organism's offspring. Over a long time the cumulative effect of the inheritance of characteristics thus acquired can even result in a new species. The most oft-cited Lamarckian exemplar is the giraffe, whose long neck is the putative evidence of such a process over many generations.[9]

In the American context, the scientist who most forcefully argued for acquired characteristics' continuing importance was Edward Drinker Cope (1840–1897). Cope's greatest fame was as a paleontologist, but his interests ranged over a much broader intellectual terrain. His first theoretical writings on evolution appeared in 1868, and were republished along with subsequent essays as *The Origins of the Fittest* in 1887. Cope's writings sought to carve out a domain distinct from Darwinism in the following way. Cope

7. Ibid., 39.

8. Bowler, *Mendelian Revolution*, 21–45. The history of the inheritance of acquired characteristics stretched back at least as far as Aristotle. This history, with copious illustrative quotations from several centuries of primary literature, was ably surveyed by Zirkle, "Early History," 91–119.

9. Bowler, *Mendelian Revolution*, 37–38.

was dissatisfied that the concept of the survival of the fittest did not press far enough into the origins of fitness in the structures of organisms and their offspring. Cope was fascinated with the notion (previously articulated by Lamarck) that the gains of development made in the lifetime of an organism are conserved and passed along to its progeny. Cope saw himself as founding a school of thought which served as a more comprehensive alternative to Darwin. He wrote of this movement: "Another school of evolutionists have therefore maintained that such structures are due to the effect of effort, i.e., stimulus or use, exerted by the living being on its own body, and that the design thus displayed is an expression of the intelligence at some time possessed by itself."[10]

Cope put great stock in the experiences of humans and other creatures creating a type of memory from both conscious acts and reflection on unconscious acts. Life impresses millions of impressions on an organism over the lifetime, most of which are unconscious in character, and are "stored, each in its appropriate place, to be sprung into consciousness" when needed.[11] Cope emphasized repetitious acts that once learned become so ingrained and automatic that they can be passed along to the progeny. He postulated: "This is true of such voluntary acts as we perform most readily automatically, and such as might be supposed to be most probably acquired by hereditary transmission, as for instance speaking."[12] As a corollary to this view, abilities could be lost over time via disuse. Cope acknowledged "the frequently unyielding character of the structures of adult animals" could pose a problem for his theory. In this context he confidently countered: "It is well known that the mental characteristics of the father are transmitted through the spermatozooid." Thus the mental life of a parent (a phenomenon he also ascribed to the maternal influence) could impress itself on the offspring. Citing phantom limbs in amputees as evidence, Cope inferred that if the long-standing impression of the organ on the mind had produced an unconscious picture of its presence, this memory-like influence on the brain cell could be passed along in procreation. Cope theorized "that the energies communicated to the embryo by the spermatazooid and ovum will partake of the character of the memory thus created."[13]

The broader philosophical payoff for Cope was a confirmation of his own idealist tradition in contrast to the more materialist emphasis of Darwin and his followers. "If the law of modification of structure by use and

10. Cope, *Origin of the Fittest*, 390.

11. Ibid., 391.

12. Ibid., 393.

13. Ibid., 407–8.

effort be true," he pled, "it is evident that consciousness or sensibility must play an important part in evolution." Even among animals, habits "may be looked on as movements acquired in consciousness, and become automatic through frequent repetition," and then "transmitted to the succeeding generation" so that actions of the offspring will become automatic and unconscious.[14] Cope claimed to have arrived at the hypothesis of "the law of use and effort" independent of Lamarck, having discovered Lamarck's 1809 work only after developing his own theory.[15] That he held it to be an all-encompassing theory is evident in Cope's assertion that feelings such as pain, indifference, and pleasure (which in humans become likes, dislikes, emotions and passions), are at least present in some form across the entire biota. In his chapter on "The Origin of the Will," Cope strove to analyze the affective dimension of experience "in order to show that the feelings in their various grades are the motives of action in all animals, from the *Amoeba* to man."[16]

Cope's *Origins* concluded with remarks indicating how he envisioned the social implications of his theory. "It is true that moral character is inherited," Cope averred, "and that changes in this department for better or worse are transmitted to offspring." In Cope we find a setting of the stage for an intellectually sanctioned cultural embrace of such notions as the hereditary criminal and the hereditary alcoholic. Cope also argued that the doctrine of the survival of the fittest is not only vindicated in natural law but in moral law. Thus we find a bridge between a scientific and a theological assessment of human heredity. The groundwork for various social reformists' paternalistic efforts to control the weak of society is set in part by Cope's neo-Lamarckism. Cope's arguments were also incorporated by leading sociologist Lester Frank Ward (1841–1913) in his address defending the inheritance of acquired characteristics before the Biological Society of

14. Ibid., 413.

15. Ibid., 423. The credibility of this claim can be assessed by the reader, though historians note that Cope was a rather self-aggrandizing person. Cope is described by biographer Joseph M. Maline as "the leading theorist of the neo-Lamarckian movement in American biology." See *Dictionary of Scientific Biography* (1978) sv: "Cope, Edward Drinker," 15: 91–93. It is also important to point out that Lamarck historians are often nonplussed with variations of his views that have subsequently born his name. So for example, whereas the neo-Lamarckism of Cope seems very close to vitalism, historian Richard W. Burkhardt Jr. insists that "Despite the vitalistic ring of the phrase 'the power of life,' Lamarck was no vitalist. To the contrary, the whole of his biological thought was decidedly materialistic." Further, the idea of the inheritance of acquired characteristics did not originate with Lamarck, but was part of the discussion of biology prior to his day. See Burkhardt, *Spirit of System*, 167, 179.

16. Cope, *Origin of the Fittest*, 445.

Washington in 1891.[17] In order to see how the debate played out over time, we next examine the views of heredity held by neo-Lamarckism's leading opponent: August Weismann.

German biologist August Friedrich Leopold Weismann (1834–1914) coined the term "germ plasm," which became a favorite theme in the eugenics social reform literature in the early decades of the twentieth century. In his 1893 tome, *The Germ Plasm: A Theory of Heredity*, Weismann at times struggled to explain the blending of characteristics that make up the unique individual of whatever species the biologist is investigating. Drawing mostly on studies of plants, insects, and various mammals, Weismann only occasionally applied his theories to humans. Extensions of his theory of heredity into human heredity remained cautious and restrained in the sense that his focus was physical traits and not complex social patterns. It was later social reformers who would apply Weismann's terms to much more complex fields of human behavior and social interaction, thus extrapolating his insights far beyond their original intent.

Weismann sought to explain why children sometimes bear stronger resemblance to their grandparents than to their parents. Important for later social reforms was his explanation of the concept of reversion. Weismann stressed that in most cases reversion is merely partial and not complete. He recognized that reversion was not a matter of simple proportionality because he observed in the germ plasm a variety of determinants ("idants" in English) mixed in unequal groupings of varying power in the progeny.[18] With reference to humans, Weismann wrote: "ontogeny is hardly ever passed through without reversions occurring to one or other of the grandparents." He viewed "ids" (groupings of identical hereditary determinants passed along in the germ plasm) as suppressing one another over the generations, but that combinations of ids may preponderate over other groupings "under certain circumstances."[19] Weismann recognized that the rules governing transmissibility of characters differed between somatic cells and germ (reproductive) cells; a greater plasticity obtained of the former as compared with the latter. Weismann asserted:

> . . . [O]nly those variations are hereditary which result from the modification of several or many determinants in the germ plasm, and not those which have arisen subsequently in consequence of some influence exerted upon the cells of the body.

17. Ward, *Neo-Darwinism and neo-Lamarckism*, 60–61. Ward served as the first president of the American Sociological Association.

18. Weismann, *Germ Plasm*, 310–11.

19. Ibid., 312; on "ids" see 60–62.

In other words it follows from this theory that *somatogenic or acquired characters cannot be transmitted.*[20]

It is well known that Weismann had bred several generations of mice from mice with severed tails to show that the missing tail did not become permanent within the germ plasm. This was seen by many as refutation of the inheritance of acquired characteristics, which had been a mainstream hereditary view up until Weismann's experiments. Yet Weismann was not claiming that "external influences are incapable of producing hereditary variations." He insisted that, "on the contrary, they always give rise to such variations when they are capable of modifying the determinants of the germ plasm."[21] The big question would become what exactly are those environmental influences powerful enough to modify a germ plasm that, in most respects, stood resistant to such change? When somatic cells are altered, their changes are both temporary and non-hereditary. "But when they occur in the germ-plasm," Weismann warned, "they are transmitted to the next generation and cause corresponding *hereditary variations in the body.*"[22] Thus neither hard nor soft versions of heredity were absolute, and each admitted some influence of both nature and nurture, even if the degree of influence allowed to each varied widely across hereditary theories. At least one Lamarck biographer argues that inheritance of acquired characters was never a central theme of Lamarck's own system, and only became so at the hands of later opponents, most notably Weismann.[23] As seen in so many popularizations of scientific ideas, the nuance of original authors gets lost in the polarizing attempt to derive and promote social policy based upon their work.

Historian Daniel Pick rightly notes that the "issues of environmentalism and heredity" were not settled by Weismann and his role for the germ plasm in inheritance.[24] Even in the arena of the social reforms envisioned by eugenicists, who in the main tended to be strong hereditarians, Lamarck's notions of the plasticity of heredity continued to have an influence. "Despite Weismann's theory of the indestructibility of the germ plasm and the rediscovery of Mendelism," Pick states, "most eugenists strongly believed that 'superior' and 'inferior' acquired characteristics could be passed on directly."[25] Paradoxically the germ plasm was plastic enough to be modified

20. Ibid., 462, emphasis in the original.
21. Ibid., 462–63.
22. Ibid., 463, emphasis in the original.
23. Barthelemy-Madaule, *Lamarck*, 82.
24. Pick, *Faces of Degeneration*, 100.
25. Ibid., 101.

by those social pathologies eugenicists found most offensive, yet deterministic enough to resist improvement by any social reforms that stopped short of guaranteeing the total prevention of the procreation of the allegedly unfit.

At the turn of the century, social reformers were not hesitant to warn of all sorts of degenerative effects of maladaptive or immoral behaviors of human individuals upon the germ plasm, and thus on the human race as a whole. While those influenced by Weissmann's theories would increasingly limit the influences that could modify the germ-plasm or the genes, the theory of the inheritance of acquired characteristics exerted continuing influence on social reform thought well after Mendelism became public knowledge after 1900.

Biologist Charles Davenport on Heredity and Religion

The connection between religious themes and the emerging science of heredity and the eugenics movement is seen in bold relief in the writings of Charles Benedict Davenport (1866–1944). Davenport's biography is a well-rehearsed story in the literature on the American eugenics movement. Educated in biology at Harvard University in the 1890s, Davenport was a rising star. After a brief stint at the University of Chicago, Davenport's potential was seen by Andrew Carnegie, scion of Gilded Age industry. With a ten million dollar endowment from Carnegie, Davenport established a station for the study of evolution and its applications to human problems at Cold Springs Harbor on Long Island in 1904. Though he was a scientist of some skill, Davenport's abilities and enthusiasm as a fundraiser and social reformer took center stage. His willingness to use Mendelian notions (very loosely construed, as it turned out) to oversimplify complex personality traits, or to promote sterilization of the "unfit" became legendary, sometimes even to the consternation of other eugenicists.[26]

In 1911, Davenport wrote: "the relation of eugenics to the vast efforts put forth to ameliorate the condition of our people, especially in crowded cities, should not be forgotten."[27] After surveying efforts to improve such conditions through education, religious teaching, and public health reform, Davenport asked: "What can be done to reduce the frequency of the undesirable mental and bodily traits which are so large a burden to our population?"[28] At this stage Davenport was skeptical toward involuntary

26. Kevles, *In the Name of Eugenics*, 44–49.

27. Davenport, *Heredity*, 254. See also Zenderland, *Measuring Minds*, 226–27.

28. Davenport, *Heredity*, 255–56.

sterilization and marriage restriction for the feebleminded and criminal, though largely on practical grounds. He instead espoused segregation of the feebleminded during their period of fertility, a fairly common and conservative response among caretakers of the feebleminded at the time.[29]

In a section titled "The Salvation of the Race through Heredity," Davenport criticized "the hazy doctrine of the freedom of the will." In this context Davenport was willing to ameliorate culpability for bad social behavior. "How we respond to any stimulus depends on the nature of our protoplasm," he averred. Davenport admitted that training or formation of habits could moderately modify human responses to stimuli, but urged that this was limited "by the impressibility of the protoplasm." In a turn of phrase far kinder than what would become the norm among eugenicists, Davenport wrote: "So I do not condemn my neighbor however regrettable or dangerous he may be."[30] Drawing upon what would come to be known as positive eugenics, Davenport compared human procreation with plant and animal breeding. By using strategic cross-breeding "to increase the vigor and productivity of their stock and crops," such scientists showed the way to human improvement. Based on this analogy, Davenport assured his readers that "proper matings are the greatest means of permanently improving the human race—of saving it from imbecility, poverty, disease and immorality."[31] Salvation language had been taken from the religious arena and applied to science and social policy.

Davenport would become a tireless promoter of eugenics via print media such as numerous articles and books, and through his speaking engagements. Davenport's deeper theological convictions are not readily apparent, but his fervor and his zeal for eugenics may be likened to that of the Christian missionary. His use of biblical rhetoric was doubtless drawn in part from his father's puritanical strain of observant Congregationalism, and the long list of Congregationalist ministers in his pedigree, stretching back to England.[32]

As he spread the message of eugenics, Davenport sought common cause with a wide variety of leaders in American society, including clerical leaders. One method by which Davenport appealed to religious audiences was to portray eugenics as a religion. In this he was following in the footsteps of the founder of eugenics, Francis Galton. Neither figure evinced much respect for the intellectual mettle of traditional religion, but rather

29. Ibid., 259.

30. Ibid., 260.

31. Ibid.

32. Black, *War Against the Weak*, 32.

cynically perceived its emotional potentialities and powers of persuasion as amenable to the eugenics cause. For Davenport, religion was a motivational force, whereby the passions could be aroused for the promotion of some putative social good. Religion thus served a useful function when channeled and directed by the dictates of science.

In 1916 Davenport's blending of religion and science emerged explicitly in his address, *Eugenics as a Religion*. Davenport's speech began with some ground clearing, whereby he criticized religion for its emphasis on environment in the care of the less fortunate. Targeting the history of Christianity's involvement in charitable works among the feeble-minded, he proclaimed that religions usually "fail to regard the importance for society of inheritable racial traits." According to Davenport, eugenics made up this deficit by its primary aim, namely, "to improve social conditions by securing to the next generation the greatest proportion of persons best fitted by nature to carry on each his own share of the world's work." He contended that the highest purpose of humanity was "to develop a social order of the highest, most effective type—one in which each person born is physically fit, well endowed mentally for some kind of useful work." For Davenport, this also entailed the creation of an ideal person, who would be "temperamentally calm and cheerful and with such inhibitions as will enable him to control his instinctive reactions so as to meet the more of the community in which he lives." Seeking to rally the crowd at Battle Creek Sanitarium, the original location of his speech, Davenport posed the rhetorical question: "Do you agree that this is the highest aim of the species? Have you the instinct of the love of the race?" The answer was tantamount to a new faith: "If so, then for you eugenics may be vital and a religion that may determine your behavior."

Davenport ended his speech with an eleven-phrase "creed for the religion of eugenics." His repeated phraseology of "I believe" doubtless built to a crescendo that evoked the image of the evangelist. The "Creed" contained portentous phrases such as: "I believe in the maintenance of a high quality of hereditary traits in the nation," and "I believe in such a selection of immigrants as shall not tend to adulterate our national germ plasm with socially unfit traits." He ended with the enigmatic entendre: "I believe in doing it for the race."[33]

The question such material raises for historical reflection is: how could such assertions have been heard by a middle-class Protestant audience not merely with a straight face, but with loud applause?[34] The answer must reside

33. Davenport, *Eugenics as a Religion*, 6–8.

34. Ibid., 8. The parenthetical phrase "(loud applause)" appears, rather gratuitously, as the last phrase of the published text of Davenport's speech.

in a complex combination of assumptions about the nature of science, the nature of Christianity, and especially the *relationship* of science and religion.

The Ministry of Eugenics

The application of theories of animal husbandry to human breeding would be a bitter pill to swallow for many Protestant religionists. It would take the rhetorical finesse of princes of the pulpit to help the medicine of eugenics go down more easily. In 1913, the *American Breeders Magazine*, a central organ of the American eugenics movement, issued a challenge to America's preachers. Cognizant of the continuing influence of the clergy in an increasingly scientific age, the editors of this periodical made an appeal to ministers to jump on board the eugenics bandwagon. This editorial promotion of eugenics appealed to the intellectual, class, and race conceits of Anglo-Saxon Protestant ministers in a bid to make the proclamation of eugenics a task central to their ministerial vocation:

> The facts and principles of eugenics are to become a substantial part of the thought and every-day philosophy not only of educated people but also of the masses. And the assistance of all churchmen will be greatly needed to aid in developing the philosophy along wholesome lines, and in carrying to all the people those facts and rules of practice which will best serve the race. When it is realized that it must become almost a racial religion for those with superior heredity to multiply more rapidly, and for those with inferior heredity to multiply less rapidly, in order that the race may evolve upward, a new burden will be placed on the preacher. That the church lay hold of this new agency of eugenic righteousness and place it effectually beside education and religious teaching as the third great agency for the regeneration of the race, the preachers must study to know the facts, and must be schooled in interpreting the facts of science into the language of the masses.[35]

How would Protestant ministers respond to such a challenge? As with most questions of history, the answer is a complex one. As Christine Rosen has demonstrated, many (mostly progressive) ministers responded enthusiastically to the call of the American Eugenics Society for "eugenic sermons" to be preached in America's pulpits.[36] On the other hand, voices of discontent with eugenics would arise among theologians and religious scientists

35. "Preachers and Eugenics," 63–64.
36. See Rosen, *Preaching Eugenics*, 135–37.

with qualms about eugenics. Indeed, refinements of the science of genetics, coupled with a renewed appreciation of environmental factors in human flourishing, significantly contributed to the rise of a few voices of opposition to eugenic applications of scientific theories of heredity. One biologist in a religious institution who urged caution regarding hereditary determinism was Herbert William Conn.

A Wesleyan Biologist and the Power of Social Inheritance

Situated well for the exploration of the intersection of scientific and theological dimensions of human heredity was Herbert William Conn (1859–1917). Conn was an instructor in biology at the Wesleyan University of Connecticut, and full professor there from 1888 until his death in 1917. Upon receiving his PhD in invertebrate zoology from Johns Hopkins University in 1884, Conn became a professor in the Department of Natural History at Wesleyan University in Connecticut in 1884. Conn established a new department of Biology at Wesleyan, and spent his career as the head of the department until his death in 1917.[37] Most of his career blended the study of bacteria with public health applications, especially aimed at making dairy products safer for the public. Conn was a co-founder of the American Society of Bacteriologists (later the American Society of Microbiology), and served as its president in 1901.[38]

Conn's 1913 article, "Eugenics versus Social Heredity" opened with the anecdote in the public press that, while human babies in the inner cities were going for two dollars, some dogs were going for one hundred dollars. Eugenicists boasted that this was because dogs were the outcome of selective breeding, while humans were not, and this disparity constituted a debating point in favor of selective breeding for humans.[39]

The burden of Conn's essay was to argue that such a facile reading of human inheritance failed to distinguish humanity from animals on an issue of utmost importance. Whereas humans were analogous to animals in their *organic inheritance*, they were utterly distinct in their *social inheritance*. By this time, Conn was well aware that "the inheritance of acquired characteristics" was highly problematic from a scientific point of view. But

37. From the biographical essay authored by his son, H. J. Conn, "Professor Herbert William Conn, and the Founding of the Society," 288.

38. Ibid., supplemented by e-mail correspondence with American Society for Microbiology archivist Jeffrey Karr, June 2010.

he pointed out that the "noninheritance of acquired characteristics" applied only to *organic* and not to *social* aspects of inheritance. "When we come to look carefully into the question," Conn insisted, "we find that those possessions which we have inherited by the laws of social inheritance are perhaps even of more importance than those which we have inherited by laws of organic heredity; and, further, that they are as sure, if not surer, in their results."[40] Conn contrasted the hypothetically fecund citizen of 1913 who had sired twenty children over against George Washington, who had sired none. "But when we make such a comparison it becomes evident at once that the development of human civilization is controlled by a different set of laws from that which governs the evolution of animals."[41] For Conn, social inheritance clearly outweighed organic inheritance.

Conn insisted that many traits identified as organic in origin turn out to be "impressed upon us by the series of forces which we have here called social inheritance." Individuals derive their moral standards, for example, mainly from their environment and upbringing. Conn pointed out that "the laws of heredity," those traits passed down from parent to child, "are indeed immensely important; it is these with which eugenics is concerned and which it is trying to improve." Significantly for Conn, the second set of forces in a person's life was to be gained from society, for "these mold the attributes that are received from the parents, turning the possibilities of the babe into the actualities of the adult." Conn viewed social inheritance as "not the lesser, but probably the greater of the two."[42]

Many voices in 1913 were warning of the "race suicide" that would befall (especially Nordic, white) citizens as a result of careless and "dysgenic" marriages and procreational practices. For Conn, this rhetoric was overblown, and an evidence of an unwarranted emphasis upon organic over social inheritance. Conn urged his readers to remember that "the individual can count in evolution quite independently of his offspring," and as a result of individual achievement, "the problem of race suicide becomes less serious and at all events assumes quite a different aspect." Human beings can impress their influence upon the human race through individual activity without recourse to physiological reproduction.[43] Indeed, those whose reproductive powers were being swamped out by others were only showing that they had failed to be socially adaptive. Natural selection became cultural selection—those who had lost the ability to compete culturally would

40. H.W. Conn, "Eugenics" 708–12.

41. Ibid., 711.

42. Ibid., 714–15.

43. Ibid., 717.

be weeded out, and "the next generation will be placed on a higher plane because it has inherited the teachings of society."[44] Superior ideas stood analogous to superior genes in the competitive arena of social evolution. "Even though our own family or our own race may show itself to be unfitted for carrying on to the future the civilization that we have produced," Conn reassured his readers, "our influence upon the future will be none the less sure." In a remarkably egalitarian voice, Conn mused that "the races to take our place will be upon a higher plane because of the inheritance that we transmit to them by the laws of social heredity."[45]

By the following year, Conn had published his argument in expanded form in the book *Social Heredity and Social Evolution: The Other Side of Eugenics*. In a section warning against the "unfortunate results" of eugenics teaching (most notably creating an ethos of hopelessness and contributing to the neglect of "social heredity") Conn expressed skepticism toward the growing trends of marriage restriction as a path to a eugenically fitter society. "It is not possible to expect that marriages will be determined by fitness," he cautioned," nor that human breeding will ever be controlled as is the breeding of domestic animals." In a wry and realistic moment, Conn predicted succinctly: "Marriage will continue to be determined by passion and accident rather than by fitness."[46] Conn went on to argue that while biological inheritance was important, social inheritance set humanity decisively apart from the rest of the animal kingdom.[47] The rest of the book laid out in painstaking detail the varied dimensions of social heredity, drawing upon both biology and the comparative study of human cultures and culture-formation processes. Conn critiqued the popular hereditarian determinism permeating eugenic studies of families such as the Jukes and the Kallikaks.

Eugenicists loved to contrast the Jukes and Kallikaks (pseudonymous putatively dysgenic family lines) with the descendants of New England theologian Jonathan Edwards. Yet for Conn, the "conditions of vice and criminality" of the former were key to the long heritage of criminality in such family lines. If the environments of the Jukes family and the Edwards family had been exchanged, Conn averred, "it is safe to say that there would not have been as many criminals in the Jukes family nor as many college presidents or other men of note in the Edwards family."[48]

44. Ibid., 718.

45. Ibid., 719.

46. Herbert William Conn, *Social Heredity*, 12–13.

47. Ibid., 16, 287, 339–40.

48. Ibid., 295.

Conn's theological tradition is more implicit than explicit in *Social Heredity and Social Evolution*. Yet some of these implicit principles are rooted in Christian and Wesleyan worldview considerations. For example, Conn maintained the conviction that humans are distinct from other animals, primarily through the passing on of wisdom by many conduits (family rites, education, etc.) under the general rubric of social evolution. "Although in the organic evolution of animals *nature* rather than *nurture* has been the predominant force," Conn concluded, "in human social evolution *nurture* rather than *nature* has stood foremost." Eschewing the pessimism of the heralds of race suicide and negative eugenics solutions, Conn ended his book with the theological virtue known as hope. "It is not what we are born, but what we become after birth that makes us men: It is not the power of babes, but what civilization makes of those powers that constitutes the essence of mankind." Reaching an eschatological climax, Conn ended with the confident declaration: "The future is full of hope."[49]

The Wesleyan tradition had long valued social reform as part of its ethos growing out of a theology of perfectionism. Critical to understanding John Wesley's long-term impact upon Anglo-American culture is the recognition that his theology gave rise to remarkable strides in social reform in both England and the eventual United States.[50] Wesleyan scholar John L. Peters has remarked on the social ramifications of this theology, noting that "Wesley's idea of holiness was a concept deeply rooted and widely spread. Personal salvation was focal, but in the last analysis he knew no holiness but 'social holiness.'"[51] Wesley was keenly aware of human sinfulness, and espoused a version of the doctrine of original sin more rigorous than many of his religious heirs. Yet Peters notes that, "he nevertheless refused the arbitrary dictum that the grace of God can go only so far and no further in its redemptive efficacy."[52] Methodist historian Gerald Kennedy wrote: "If you ask a man who knows the Protestant churches to define a distinguishing mark of Methodism, he will likely speak of its social concern. We inherited this from our founder, who could find no holiness in the Bible that was not social." Kennedy pointed out the impact of the Methodist Social Creed of 1908 on similar statements by the pan-Protestant ecumenical movement's

---

49. Ibid., 343–44.

50. See Hamilton, "Church as a Universal Reform Society," 52; and T. Smith, *Revivalism & Social Reform*, 103–113; 148–62. Wesley's efforts to "feed the hungry, to clothe the naked, to employ the poor, and to visit the sick" by methodical organizing of such efforts via his "societies" are noted in Tuttle, *John Wesley*, 283.

51. Peters, *Christian Perfection*, 183.

52. Ibid., 185.

Federal Council of Churches.[53] Thus, the importance of nurture loomed large for theological, as well as for scientific, reasons to Herbert William Conn of the Wesleyan University of Connecticut.

Conclusion

If anything, the history of heredity in the late modern era can serve as a cautionary tale about the damage done by taking the painstaking researches of the laboratory and applying them in a rushed, slipshod, decontextualized or politicized manner to social policy applied to broad groupings of human persons. By the late nineteenth century, scientific notions of heredity and environment were routinely mined by social reformers as justification for a host of eugenic social reforms. Social reformers were drawn from a wide range of disciplines, including clergy, biologists, psychologists, sociologists, penologists, and government functionaries. Some of these social reformers used religious or theological language along with scientific language to advance eugenics agendas in ways later generations would rue. Theological or religious ideas and ideals played various roles in the story of human heredity and its social applications, roles only now beginning to be appreciated. One key phrase that became commonplace was "the stigmata of degeneration." An exploration of that term and its distortive impact is the next stage on our narrative journey.

53. In Bucke, *History of American Methodism* 3:255; cf. Norwood, *The Story of American Methodism*, 354.

CHAPTER 3

"Stigmata of Degeneration": The Religious Rhetoric of Eugenics

It is supposed by many that criminals are readily recognized by certain physical stigmata. While undoubtedly many prisoners show these characteristics and corresponding physical defects, the great majority do not.

—Joseph F. Scott, 1907[1]

Introduction

JOHN HARVEY KELLOGG (1852–1943), cereal magnate and health advocate extraordinaire, edited for several decades the popular magazine *Good Health*. The pages of *Good Health* exemplified the tension between a confidence in the superiority of the Anglo-Saxon race, and fears about the erosion of the vigor of that race. Kellogg devoted much attention to alarming statements by British intellectuals concerning this problem. He also offered his own concurrence in that assessment. In 1904, he wrote: "Even a casual visitor to London must be struck with the great number of inferior, deteriorated looking people whom he meets upon the streets. This great center of civilization seems to be also a center of human degeneracy."[2]

Physiognomics, or the determining of personality traits based on facial features, has been a source of fascination for elites since at least the time of Aristotle.[3] The great peripatetic philosopher wrote: "The physiognomist

1. Scott, "Prison and Police Administration Report," 86.

2. Kellogg, "Deterioration in Great Britain," 332; see also Kellogg, "The Race Growing Old," 668; Kellogg, "Recent Facts Regarding the Growing Prevalence of Race Degeneracy," 123; and Kellogg, "Mendel's Law of Heredity and Race Degeneration," 737.

3. Hassin and Trope, "Facing Faces," 837–52.

draws his data from movements, shapes, and colours, and from habits as appearing in the face, from the growth of hair, from the smoothness of the skin, from voice, from the condition of the flesh, from parts of the body, and from the general character of the body."[4] As seen above, the founder of the eugenics movement, Francis Galton, promoted the use of physiognomy in assessing underlying character traits in persons of the modern era. At the opening of the twentieth century, this longstanding practice of judging a book by its cover came to be regarded as an exact human science.

Comments about the appearance of those on the ostensibly lower rungs of the social ladder betook of a rhetoric that can be jarring to the reader's ears today. In the Gilded Age and Progressive Era elite sources, the language tended toward the routine dehumanization of those regarded as inadequate for full inclusion in society. It was, after all, a modern society many assumed to be moving in a steadily progressive and upward direction. Such persons were seen as a drag on society's move toward ever greater heights of civilization, a key term universally assumed by elites to be an unqualified good toward which all should strive. This chapter explores a particular concept routinely applied by eugenicists to the allegedly unfit. This trope is important because it, like "degeneration," bespoke a merging of religious and scientific sentiments that masked a prejudicial position toward various less fortunate citizens in American society. No figure refined and honed the term "stigmata of degeneration" more rigorously than did Arthur Macdonald.

Arthur MacDonald and Criminology

Arthur Macdonald (1856–1936), a prominent criminologist in the Progressive Era, gave testimony before the U.S. Senate Subcommittee on Juvenile Crime and Reformation on March 6, 1908. Document 532, published by the Government Printing Office on May 21 of that same year, illustrates the reach of degeneration theory into the realm of social policy in the early twentieth century. The rhetoric of the document indicates the fear of those elites bent on the social control of the "lower" echelon of the populace. In the Progressive Era, few received more vitriolic treatment than those who seemed regressive, or resistant to social control, and thus whom elites deemed to be defective. The wording of the opening paragraph of the bill mirrors a widespread tendency to lump together with criminals those who were poor or struggled with intellectual disabilities:

4. Aristotle, *Minor Works*, 93.

> *A BILL to establish a laboratory for the study of the criminal, pauper, and defective classes. Be it enacted by the Senate and House of Representatives of the United States of America in Congress assembled,* That there shall be established at Washington, in the District of Columbia, a laboratory for the study of the abnormal classes, and the work shall include not only laboratory investigations, but also the collection of sociological and pathological data, especially such as may be found in institutions for the criminal, pauper, and defective classes. . . .[5]

Here MacDonald used "stigmata of degeneration" as a quasi-scientific term of identification and applied it to numerous American citizens who, with hindsight, were largely undeserving of policies aimed at controlling intimate details of their lives. This source illustrates how one generation's science-based social policy looks, to posterity, like prejudicial governmental overreach.

MacDonald was the author of at least 25 books in the fields of sociology and criminology between 1893 and 1918.[6] A student of famed educational psychologist G. Stanley Hall at Johns Hopkins University in the 1880s, MacDonald likely also interacted with the Italian father of criminology, Cesare Lombroso, among other European scholars in that nascent field. According to Bremner and Barnard, his effort to establish a scientific lab for the study of young delinquents did not meet with success.[7] According to a leading historian of criminology, Macdonald's 1893 tome *Criminology* was the first U.S.-published book-length text on the causation of criminality, though his thought was largely derivative, scientifically naïve, and built on scientific racist assumptions.[8] Clearly he had connections among the powerful in Washington DC, as more than one of his laborious works was published by the Government Printing Office. This was not enough leverage however to bring to fruition his dream of a publicly-funded laboratory dedicated to the study of defective delinquents. Still, his development of the concept of the "stigmata of degeneration" serves as a window on the kind of totalizing discourse eugenicists employed so as to portray eugenics policies as necessary in the eyes of the average citizen.

---

5. MacDonald, *Juvenile Crime*, 7.

6. See Hathi Trust listing at URL: http://onlinebooks.library.upenn.edu/webbin/book/lookupname?key=MacDonald%2C%20Arthur%2C%201856–1936.

7. Bremner and Barnard, *Children and Youth in America*, 2:562–3.

8. Rafter, *Origins of Criminology*, 188.

Science and Social Control

In his opening remarks before the Senate subcommittee, Macdonald stated that for many years he had been laboring to produce legislation, both in the U.S. and globally, "for founding laboratories to find the best methods of preventing crime, pauperism, and defectiveness, the most constant, most costly, and greatest enemies of all government."[9] Here American citizens, due to problems rooted in a mixture of congenital, environmental, and behavioral factors, MacDonald forthrightly described as "the greatest enemies" of their government. While a rare case might be made for the legitimacy of such rhetoric toward those who truly committed crimes against the state, were tried by a jury of their peers, and convicted via due process, Macdonald conflated the poor and the mentally handicapped with criminals. Such conflation was routine in Progressive Era social reform literature. A kind of rhetorical distancing of elites imbued with governmental power from ordinary citizens not guilty of any substantive wrongdoing helped to legitimate draconian social policies aimed at profound restrictions on the freedoms of said citizens. The growing cultural authority of science was used in this expansion of social control, as I illustrate below.

His elocution in March of 1908 was not MacDonald's first attempt at persuading the U.S. Senate. His testimony two years earlier to the 58th Congress on the same themes had previously been published several years earlier (1893) as *Man and Abnormal Man*. Through the work of criminologists such as MacDonald, the political pressure had been building for well over a decade for government action. MacDonald asserted "some ninety learned bodies in science, medicine, law, and religion for the last six or seven years have been asking Congress for the enactment of this or some similar bill into a law." He provided names of many prominent organizations favoring the legislation, including The American Medical Association, The American Bar Association, and Protestant denominations ranging from Presbyterian to Baptist to Congregational to Unitarian.[10]

In the ensuing marshalling of studies and statistics to validate the need for such laboratories, MacDonald assured the subcommittee that "hopeless cases are very few," and thus reform work had a chance of success. Still, he noted that some children "are born with feeble moral tendencies to such a degree that reformation is impossible. . . ." Further, he emphasized that "these are frequently cases of moral degeneracy with strong hereditary taint. In some instances it would be as difficult to reform their characters as to

---

9 MacDonald, *Juvenile Crime*, 7.

10. Ibid., 8–9.

change the shape of their heads."[11] This allusion introduces the connections of degeneration theory to an earlier science important to the opening decades of criminology: phrenology, or the studying of crania to determine mental or criminal tendencies.[12] The term "phrenology" was falling out of favor among biologists by the early twentieth century, but the concept of seeing visible signs of underlying disability on the faces or skulls of institutionalized persons remained commonplace in the Progressive Era. The moniker for this reading of defect in the bodies of the underclass, widespread in eugenics literature, was "stigmata of degeneration." This theme, forming the subtitle of MacDonald's published testimony, received intensive focus for thirty pages of fine print.

MacDonald's use of the phrase was not idiosyncratic. For instance, Martin Barr, whose 1904 book *Mental Defectives* became influential among mental health experts in the early twentieth century, wrote:

> Feeble-mindedness, including idiocy and imbecility, is defect either mental or moral or both, usually associated with certain physical stigmata of degeneration. Although incurable, its lesser forms may be susceptible of amelioration and of modification, just in proportion as they have been superinduced by causes congenital or accidental.[13]

Barr, superintendent of the Pennsylvania State Training School, would author the 1905 legislative language aimed to persuade Pennsylvania legislators to legalize involuntary sterilization for those "idiots" deemed incapable of improvement and whom physicians declared unfit to procreate. The governor vetoed the bill, and Pennsylvania never went on to legalize the procedure.[14] However, by 1940 some 30 states did eventually legalize involuntarily sterilization, a fact for which in the early years of the twenty-first century some states have been officially apologizing. For a brief time those favoring segregation of those thus stigmatized opposed those who favored sterilization as the preferred method for reining in the

11. Ibid., 15.

12. See Gould, *Mismeasure of Man*, 151–72; cf. Rafter, *Origins of Criminology*, 190, for evidence of MacDonald's obsession with cranial and facial features.

13. See Barr, *Mental Defectives*, 23. One early history of the involuntary sterilization movement helpfully surveys details of the passage of sterilization laws chronologically and by state. See Landman, *Human Sterilization*, 52–93.

14. Largent, *Breeding Contempt*, 69–70, cf. 11, 68, 86. Largent's survey of the early years of coerced sterilization legislation is detailed. As seen in his analysis, many pieces of the pro-sterilization literature built up prior to such legislative efforts made prominent usage of the language of degeneration theory.

procreation of the feebleminded. Yet within a decade, the sterilization method had carried the day.[15]

Stigmata of Degeneration Proliferating

In Arthur MacDonald's testimony on the stigmata of degeneration we find many echoes of the fear of going backward, or reversion, commonly dreaded among the progressives. That fear took scientific form in terms of atavism, or the "throw-back" that appeared not merely in science, but in the fiction of the era as well. MacDonald defined degeneration as "an essence, aptitude or tendency which hinders development (mental, moral or physical) favorable to the species and tends toward diseases, which lead to the dissolution of the species or offspring."[16]

Under the heading "Physical Stigmata of Degeneration," MacDonald surveyed the scientific literature in physiology. With paragraphs on cranial, ear, eye, nose, mouth, palate, tongue, teeth, trunk, hands, and genitals, Mac-Donald sought to leave no physical evidence of degeneration unexplored.[17] The topic "Functional Stigmata" also received its due, broadly including anomalies of speech, movement, genital function, and anomalies of sensory and vasomotor systems.[18]

Under the heading "Mental Stigmata of Degeneration" MacDonald noted debates over the causes and classifications of the insane. He acknowledged both hereditary insanity and accidental insanity (i.e., mental disease) in an otherwise normal person. Yet here he set forth many qualifications that tacitly undermined the usefulness or probity of degeneration theory. For example, in a subsection treating of "those predisposed with degeneracy," MacDonald tried to distinguish between those who are predisposed with degeneracy but do not manifest it from those predisposed to degeneracy that does in fact become manifest. Thus a bifurcation of potential and actual degeneracy was introduced. Still, even for those who merely had it potentially, MacDonald averred that anomalies may take mental form such as "sentiment, intelligence, instincts or inclinations," or may take physical form, and these are added to "the concomitant mental anomalies." Thus MacDonald lumped physical and mental features together. "All these stigmata are permanent," MacDonald declared, "and born with the individual, and continue until death." MacDonald admitted the environmental factors

15. Dowbiggin, *Keeping America Sane*, viii.

16. MacDonald, *Juvenile Crime*, 270.

17. Ibid., 271–80.

18. Ibid., 282–83.

at work in degeneration, such as "bad mental, physical, moral, or social surroundings" that "can easily develop this degenerative trait." Rather chauvinistically, he urged that "even physiological moments, such as puberty, menopause, menstruation, and pregnancy may make degenerative taint manifest." Without a sense of irony or cognitive dissonance, MacDonald made the following claim: "Some of these degenerates may have brilliant minds, but they are without equilibrium; they may be eccentric, bizarre, peculiar, and original. They are superior degenerates."[19] One must allow for all sorts of degenerates, it seems, when testifying before the U.S. Senate, even in 1908.

Under "Moral Stigmata of Degeneration" MacDonald offered a stunning laundry-list of sins, and the reader wonders if the Senators themselves squirmed in their seats when hearing the list:

> Any act is a moral stigma of degeneration in which there is a permanent tendency or inclination: To indulge in any form of vice, dissoluteness, depravity, profligacy, vileness, or loathsomeness; to use any form of deception, as lying, fraud, trickery, imposture, etc.; to any kind of meanness, villainy, baseness, etc.; to extreme selfishness, self-love, egotism, stinginess, covetousness, etc.; to cowardice, poltroonery, extreme distrust or suspiciousness, etc.; to any form of cruelty, brutality, inhumanity, etc.; to any form of vulgarity, coarseness, etc.; to any form of malice, hatefulness, ill-will, revenge, etc.; to laziness, indolence, listlessness, dilatoriness, etc.; to ostentation, display, pomposity, vanity, or arrogance; to frivolity, silliness, giddiness, etc.; to run into debt, insolvency, etc.; to wastefulness, extravagance, etc.; to uncleanliness, filthiness, etc.[20]

Conclusion

This list today seems to read like the biography of Frank Unger on the television series *House of Cards*, but more substantively, it represents the sheer scope of moral intrusiveness of the governing vision represented by MacDonald and many Progressive Era social reformers. If ever there was a modernist totalizing discourse, it was surely represented in such a tableau. Virtually every human being is by these terms a "degenerate" and bears "the stigmata of degeneration." Elites however often see flaws only in the other, not in themselves. Degeneration theory served as the nexus

19. Ibid., 284–85.
20. Ibid., 290–91.

between scientific and moral/religious rhetoric toward those on the margins of society. Such rhetoric would soon expand beyond the assessment of occasional individuals, and encompass social groupings of the most intimate sort. Language that demeaned persons based on their appearance took an especially insidious form in the early twentieth century known as the "Ugly Laws," whereby many municipalities sought to hide the poor and homeless from public view.[21] Thus the rhetoric explored in these pages carried an impact beyond mere academic interest. Social policies of bodily control were formulated based on judgments of character measured according to surface features of American citizens. In the next chapter we shall see that the notion of self-evident degeneration was a theme relating not merely to individuals, but to entire families and lineages. Proving this became the burden of sociologically trained social reformers devoted to the creation of a new genre, the eugenic family studies literature.

21. Schweik, *The Ugly Laws*, 89–107.

CHAPTER 4

Eugenic Family Studies, Science, and Religion

The cradle is God's purest shrine:
At this fair fount of life—
Hush here, O world, your strife!—
Bow with veiled eyes, and call divine
The mother crowned as wife.

—Minot J. Savage, Unitarian Minister, 1902[1]

Introduction

THE PROGRESSIVE ERA MARKED a time when the infant disciplines of genetics and sociology converged with the ancient role of minister in social
reform efforts to exert social control over the institutions of marriage and
the family. Central to this development was the rise of the popularity and
brief scientific ascendancy of the eugenics movement in the early twentieth century. This chapter explores the conflation of scientific and religious
rhetoric during the period of the late nineteenth and the early twentieth
century in the historical genre known as family studies literature. An elite
emphasis on heredity over environment led to social policies corrosive of
the vital boundary between the family life of the poor and the intrusive
meddlesomeness of self-appointed newly minted experts in social reform.

Oscar McCulloch and the Tribe of Ishmael

The Reverend Oscar McCulloch (1843–1891) was an early pioneer in the
field work foundational to the rise of sociology as an academic discipline. He

1. Savage, *Men and Women*, 64.

was also a local Congregational minister deeply committed to social reform and charity in Indianapolis during his pastorate at Plymouth Church.[2] He built on the work of Richard Dugdale, who in the 1877 published the first major work in the family studies literature, *The Jukes: A Study in Crime, Pauperism, Disease and Heredity*. Dugdale had set the basic parameters for the study of problematic families, providing McCulloch with an example of a more scientific approach to charity.

McCulloch's classic study of the pseudonymous "Tribe of Ishmael" serves as an early window into the genre known as the eugenic family study. The eugenics movement of the late nineteenth and early twentieth century evinced an obsession with families of "inferior stock" or of those deemed "unfit" to continue engaging in unfettered procreation. McCulloch described his first encounter with this extended family in 1877. He recounted discovering, through public records, a "pauper history of several generations," whose intermarriage with others had formed "a pauper ganglion of several hundreds."[3] McCulloch and his Charity Organization Society assistants charted the family profiles of some thirty families to form the "Tribe of Ishmael."

After laying out a litany of crimes, diseases, and sexual dysfunctions among the related families of this problematic lineage, McCulloch reached his conclusions. First he noted "this is a study in family degeneration" characterized by "parasitism, or social degradation." Secondly, various forms of unchastity were habitually practiced by the tribe, as evidenced in prostitution, illegitimacy, as well as "incests, and relations lower than the animals go." McCulloch ascribed such behaviors to deficiencies of the environments in which child-rearing occurred, notably overcrowding, indecency, and lack of cleanliness. Thirdly, McCulloch pointed to the "force of heredity." Fourthly, he blamed an indulgent society for offering the tribe of Ishmael public relief beginning in 1840, including poorhouses funded by public resources as well as private benevolence. These efforts he even labeled harshly as "the alms of cruel-kind people" due to their ostensible perpetuation of the conditions leading to familial misery.[4] Such an emphasis on both hereditary and environmental factors would undergo a shift over the next three decades in the writings of eugenicists toward an increasing emphasis on heredity. Hereditary or inborn tendencies became first a refrain and then a constant drumbeat as eugenics burgeoned into a movement bent on preventing the

2. "Rev. Oscar C. McCulloch Papers," no pages, http://www.in.gov/library/fa_index/fa_by_letter/m/l363.html. Accessed 20 October, 2015.

3. McCulloch, *Tribe of Ishmael*, 2. For more details on McCulloch, see Deutsch, *Inventing*, 19–48.

4. McCulloch, *Tribe of Ishmael*, 6–8.

procreation of a long list of so-called "defective" citizens. The sociological group who took the brunt of such development was comprised of persons increasingly deemed "the feeble-minded."

H. H. Goddard and the Kallikak Family

No social reformer was more associated with theorizing the causes and the treatment of feebleminded persons in the Progressive Era than Henry Herbert Goddard of the Training School for Feebleminded Boys and Girls at Vineland, New Jersey. His 1912 *The Kallikak Family: A Study in the Heredity of Feeblemindedness* set forth the theoretical and practical problem posed by the cognitively disabled and the familial matrices in which they were enmeshed.

The Kallikaks begins innocently with the story of a young girl, Deborah, who had come to the home for the feeble-minded at Vineland about 1898. The chronicle turns quickly to Deborah's progress reports, kept in detail by her minders, and excerpted briefly for the reader on a month by month basis over the course of several years. The dispassionate modern reader notes even with the selectivity of reportage here that Deborah seems to be a fairly normal (if a bit slow or unmotivated) young female living in an institutional setting. Yet the author's summary of Deborah's mental state intoned this description of her condition: "This is a typical illustration of the mentality of a high-grade feeble-minded person, the moron, the delinquent, the kind of girl or woman that fills our reformatories."

Today "moron" (Greek *moronos,* literally "fool") is a commonplace slur against another's intelligence in American slang. However, in the Progressive Era the term was believed to represent a scientific range of intelligence, on a differentiated scale along with "idiot" and "imbecile." *Idiots* came to refer to those with a mental age of 2 years or less, *imbeciles* with a mental age of 3–7, and the 8–12 range was reserved for *morons.*[5] Later gradations were introduced, such as "low grade" and "high grade" morons. Here Deborah became less an individual and more a symbol, as the assessment continued: "They are wayward, they get into all sorts of trouble and difficulties, sexually and otherwise, and yet we have become accustomed to account for their defects on the basis of viciousness, environment, or ignorance."[6] Goddard's account admitted an increasing role for hereditary defectiveness in the explanation of girls like Deborah.

---

5. Goddard himself was instrumental in developing the terminology, see Trent, *Inventing the Feeble Mind,* 155–66; cf. Jacobs, "Care of the Mentally Retarded," 1343–48.

6. Goddard, *Kallikaks,* 35; cf. Zenderland, *Measuring Minds,* 118, 279.

Drawing upon the notions of hard heredity and degeneration theory, Goddard and his researchers went on a journey to find the fountainhead of Deborah's alleged mental defect. They ended up identifying Martin Kallikak, the girl's great-great-grandfather. As a soldier in the Revolutionary War, he had indulged a dalliance with a feeble-minded girl from a tavern frequented by militiamen. From there, Goddard and his researchers found a branch of the family tree comprised of persons with a whole host of social, physical, and mental problems. The scientific dispassion which often characterizes the tome often gives way to emotional language: "The surprise and horror of it all was that no matter where we traced them, whether in the prosperous rural district, in the city slums to which some had drifted, or in the more remote mountain regions . . . an appalling amount of defectives was everywhere found."[7] The wayward ancestor after leaving the Revolutionary army went forth and "married a respectable girl of good family" through whom he sired a legitimate lineage described as "another line of descendants of radically different character." By 1912 the count of this upstanding side of the family was 469 of direct descent, of whom only three were found to be "somewhat degenerate, but . . . not defective."[8]

By contrast, the account refers to the entire other branch of the family tree as "the degenerate branch," and regales the reader with details of their socially problematic history. Goddard lamented: "Again, eight of the descendants of the degenerate Kallikak branch were keepers of houses of ill fame, and that in spite of the fact that they mostly lived in a rural community where such places do not flourish as they do in large cities."[9]

By the end of *The Kallikaks*, Goddard, the son of missionary Quakers, engaged in religious rhetoric to drive home his opinions about the social reforms needed. First, he cast *paterfamilias* Martin Kallikak Sr. as a negative example regarding sexual activity outside marriage. Acknowledging that "sowing wild oats" was an activity among young males of his own current generation as well, Goddard went further and interwove biblical morality with degeneration theory. "Undoubtedly, it was only looked upon as a sin because it was a violation of the moral law," Goddard intoned. "The real sin of peopling the world with a race of defective degenerates who would probably commit his sin a thousand times over," he railed, "was doubtless not perceived or realized." With hindsight a powerful rhetorical weapon, Goddard added: "It is only after the lapse of six generations that we are

---

7. Goddard, *Kallikaks*, 16.

8. Ibid., 29.

9. Ibid., 68.

able to look back, count up and see the havoc that was wrought by that one thoughtless act."[10]

The Kallikaks was a highly influential monograph that fomented public fears of the "menace of the feeble-minded." A few years later disappointing scores on Army intelligence tests only inscribed this fear more deeply into the public consciousness.[11] It also shifted attention from individual assessments of intelligence and toward the search for familial and intergenerational causes of mental or moral defectiveness. Eugenic family studies began to assume the aura of scientific probity, even though prejudicial strains recur frequently in their pages.

Arthur Estabrook, Charles Davenport, and the Nam Family

Cold Springs Harbor in New York became the epicenter of family studies research under the auspices of the Eugenics Record Office. Armies of earnest young sociologists were deployed across the heartland by eugenics leaders such as Arthur H. Estabrook (1885–1973) and Charles B. Davenport. Their mission was to ferret out problematic family lines so that the public could be warned of their prodigious procreative proclivities, and their tales of woe could serve to buttress social policies aimed at their reproductive restriction.

One such study, published in 1912, was *The Nam Family: A Study in Cacogenics*. The term "cacogenics" was coined by eugenicists to indicate "evil births" as the foil against which eugenics or "good births" could be promoted. The book recounts the study of multiple generations of inadequate stock. Funded by wealthy railroad widow Mrs. E. H. Harriman, the investigation of the Nams was carried out by Estabrook and his assistants during the period 1911–1912.[12]

Estabrook cited earlier studies of families such as the aforementioned Jukes, as well as "the Zeroes," as an established genre for the study of dysfunctional and hereditarily defective family lines. In the introductory "Early History of the Nam Family," Estabrook began with the year 1760, highlighting the union of "a roving Dutchman" and "an Indian princess"

10. Ibid., 103.

11. Kevles, *In the Name of Eugenics*, 82–83.

12. In 1910 Mrs. Harriman funded the founding of the Eugenics Record Office at Cold Spring Harbor near where Charles Davenport conducted eugenics research. This gift included 75 acres of land and a provision of $20,000 per annum for operating expenses. Her patronage of eugenics from 1910 to 1918 was an estimated half-million dollars. Kevles, *In the Name of Eugenics*, 54–55.

in Massachusetts. Early histories of the clan labeled its members as "vaga-
bonds," who among other activities, "were apt to fall into temptation and
rum." One Joseph Nam had eight children, of which five had migrated to
New York in 1800. A few of the Nams prospered due to their industry, but
"the majority, however, were ignorant, unintelligent, indolent, and alcoholic," thus failing to improve their standard of living. The verbal portrait
describes their squalid living conditions with thinly-veiled horror. "In one
place, during the winter months," the description laments, "thirty-two
people of both sexes slept together in one room." The social reformer concluded, "Such conditions as these can lead only to illegitimacy, inbreeding,
and their attending evils of pauperism and dullness."[13]

Subsequent chapters lay out detailed analyses of the moral and physical characteristics of multiple lines of the Nam family. Further chapters survey intermarriages of the Nam lineage with such pseudonymously dubbed
tribes as "The Nap Family," "The Nars," and "The Nats." In a section sum-
marizing the findings of the study of these families certain characteristic
features of a persistently problematic nature emerged for further analysis.
"Alcoholism is extraordinarily high," the narrative states, while noting: "of
the females 88 percent and of the males 90 percent are given to drinking in
excess." While acknowledging some role for environmental or nurture factors in social problems thus far identified, the author offered a section entitled "The Inheritableness of the Non-social Traits." The reader encounters
a section on indolence, as contrasted with industriousness. With both parents identified as "indolent," the resulting offspring Estabrook described as
76.5 percent "unindustrious." In an effort to apply Mendelian ratios (rather
uncritically) to vaguely defined descriptors such as "lazy" and "industrious,"
the authors noted that "laziness carries an inhibitor which is transmitted
to the offspring." For example, in a case where a lazy mother (herself the
offspring of two lazy parents) married an industrious husband, their union
yielded 9 out of 10 children afflicted with laziness, yet only one child who
was industrious. The author concluded that "such matings are eugenically
unfortunate."[14]

The next section of non-social traits focuses on alcoholism. Here the
author cited alcoholism as the cause of increased "imbecility and epilepsy."
When both parents were alcoholic this yielded an outcome of 38 percent
as either imbecile or epileptic; when only one parent was alcoholic this
dropped to 13–14 percent as imbecile or epileptic. When both parents
were identified as "temperate" the result was 20 percent imbecile and none

13. Estabrook and Davenport, *The Nam Family*, 1–2.
14. Ibid., 65–67.

epileptic. In a fascinating detail, Estabrook observed "when the mother is not alcoholic, whether the father is or not, the frequency of imbeciles in the progeny falls strikingly." The section concludes that "alcoholism in the mother affects the mental development of her children."[15]

The influence of biblical terminology, notably the King James Version, may be detected in the next section, entitled "Licentiousness." Moral and behavioral categories of analysis would usually be paramount in a discussion of an issue such as sexual sin, but under the pen of Estabrook an effort to make a moral issue into a scientific one rose to the fore. "The network of the Nam and allied families is characterized by a large amount of harlotry and prostitution; in fact they are the main anti-social acts of this locality," he declaimed. Convinced that marriage did not carry the same social weight among the Nams as it did in society generally, the author chalked the moral failings of the group up to mental incapacity. Thus "the ideals of marriage and chastity" in Nam Hollow, Estabrook cautioned, "is insufficiently recognized, largely because the mentality of the people is not capable of appreciating their importance." Heredity once again became the predominant causal explanation when the researchers encountered two sisters who were reared in "chaste strains," yielding two more chaste generations due to chaste unions. However "one daughter . . . married a man who belonged to a neuropathic strain and had among her 8 children a religious fanatic, a macrocephalic dwarf, and 4 highly erotic males."[16] Such conclusions underwrote a distraught social tale that merely one mistaken marriage could have disastrous consequences for society.

Estabrook's narrative turns next to the problem of consanguineous (blood related) unions among the Nams. He offered the shocking claim that "nearly a quarter of all matings of Nams are consanguineous." Sociological factors such as living in a valley (geographical isolation) as well as their notable "clannishness" and "their unsavory reputation" also inhibited marriage prospects for the Nams. Bleak language concludes the narrative, as Estabrook lamented that: "the consorts selected from outside are frequently quite as defective as those who select them."[17]

Anna Wendt Finlayson and the Dack Family

Another bulletin of the Eugenics Record Office family studies literature appeared in May of 1916. With a preface by Charles B. Davenport lauding the

15. Ibid., 67–68.
16. Ibid., 68.
17. Ibid., 72.

hard work of eugenics field workers, much of its wording was defensive. Noting criticisms of the work of the Eugenics Record Office, Davenport insisted that the family studies were scientifically sound, utilizing the tools of medical training and psychiatry.

Finlayson investigated the pseudonymously named Dack family, located in the west central region of the state of Pennsylvania. A few samples will suffice to explore the rhetoric of the family study and identify its scientific and religious elements. One subject, Carrie Dack, died of Tuberculosis at age 32. At 25 she had become "mentally deranged," resulting in admission to a "hospital for the insane." Further, the account notes that once she was admitted "she talked incoherently, and was excited, mostly on religious subjects." Though discharged some ten months after admission as "restored," her husband was convinced she did not seem "entirely normal mentally." Out of her four children one was admitted to the Warren State Hospital at age 30 after having trying suicide by self-immolation. Still, the field worker described her as a well-behaved complainer, who is "easily offended, self-centered and seclusive" and "is often irritable and petulant."[18] Such terminology strikes the modern reader as inexact and crudely homespun, but such verbiage achieved the level of commonplace in the genre of eugenic family studies.

One figure, Noah Dack, came in for harsher language still. Finlayson lambasted Noah as "a lazy good-for-nothing who has never done much work and at times has been supported by the township." He lived for a time with "a notorious character" by the name Maggie Rust, with whom he sired three "illegitimate children." The couple had been arrested two times on charges of "fornication and bastardy" and the constable eventually forced them to marry.[19]

The tendency of family studies literature to lapse into *ad hominem* attacks on physical features is starkly evident in the description of one Dillie McGinness, aged 70. Portrayed as living a "narrow, self-centered life" isolated from her relatives, she looked profoundly unpleasant to the eyes of the field worker. "Her facial expression reveals her character strikingly"; the author sneered, adding that "her features are suggestive of an animal, her eyes are small and bead-like and her wrinkled face is entirely lacking in humanness."[20] Quite literally therefore Finlayson used dehumanizing language to describe certain members of the Dack family.

18. Finlayson, *Dack Family*, 14–15.
19. Ibid., 20.
20. Ibid., 22.

A boy named Jack, age 15, appeared to Finlayson as naturally "lazy, shiftless, and quick-tempered." Considered mentally unsound, she noted that "he gets many queer religious ideas and talks a great deal about the invention of a perpetual motion machine which the Lord, he says, will aid him in doing." The investigator seemed here to have little appreciation for the imaginativeness that might be expected of a youth in his mid-teens.

The account *The Dack Family* as a whole is replete with references to alcohol abuse, low intelligence, and unpleasant physical and character features. Moral failings permeate the narrative like a veritable thesaurus of inadequacy. Many of the judgments Finlayson offered using the passive voice, indicating that the criticisms leveled by neighbors and townsfolk where the Dacks resided had been routinely accepted at face value. The document concludes with "two factors" that for the author "seem to lie at the bottom of the degeneracy shown by this family." The first factor was nervous instability, and the second a "lack of mental ability." The account of the Dacks concludes with dire warnings about consanguinity, cautioning that "the marriage of cousins of defective stock produces a large proportion of defective offspring."[21]

Marriage Restriction Debates

What were the social policy implications of these rather strange tales of defective families? According to eugenics encyclopedist Ruth Engs, social hygiene reformers sought legislation mandating the issuance of a "marriage health certificate" to assure the public that a prospective couple was free of venereal disease. Further, "by 1912 some type of marriage restriction had been enacted in thirty-four states or jurisdictions in the United States."[22] Ministers were frequently enlisted by the eugenics movement in an effort to screen marriages to bring about eugenically acceptable pairings. At the Race Betterment Conference of 1914, amidst a veritable who's who of eugenic movement elites as speakers, the Rev. Walter Taylor Sumner gave a speech entitled "The Health Certificate-A Safeguard Against Vicious Selection in Marriage."[23]

The results of such efforts however were decidedly mixed. In 1910 F. W. Hatch, General Superintendent of State Hospitals in Sacramento, California wrote to the state Attorney General in defense of the state "asexualization law." Part of the rationale Hatch articulated for supporting involuntary

21. Ibid., 44–45.
22. Engs, "Eugenic Marriage-Restriction Laws," 53.
23. Sumner "The Health Certificate," 509–13.

sterilization was the perceived inadequacy of efforts for marriage restriction. "Idiots, imbeciles, and degenerate criminals are prolific," he warned, and added "their defects are transmissible." After briefly surveying the effects of such laws in multiple other states, Hatch concluded: "unfortunately, matrimony is not always necessary to propagation, and the tendency of these several different laws is to restrict procreation only among the more moral and intelligent class, while the most undesirable class goes on reproducing its kind, the only difference being that illegitimacy is added to degeneracy."[24]

Historian Christine Rosen points out that geneticists began to back away from eugenical claims about the value of marriage health certificates around 1913. Further, as evidence of the unpopularity of such efforts she cites the rise of unrestricted marriages in states bordering those with stricter requirements, as well as the advent of a form of popular vaudevillian mockery skewering the idealism of the eugenicists' vision of scientific marriage.[25]

Eugenicists however remained convinced that a failure to restrict dysgenic marital unions would contribute to the degeneration of the race and the descent of future generations into ruination by imbecility and idiocy. Within a decade prominent eugenicist Havelock Ellis would plead with the readers of his manual on "love and virtue" that:

> It is not only our right, it is our duty, or rather one may say, the natural impulse of every rational and humane person, to seek that only such children may be born as will be able to go through life with a reasonable prospect that they will not be heavily handicapped by inborn defect or special liability to some incapacitating disease.[26]

On the side of positive eugenics, a 1917 marriage manual by Professor T. W. Shannon, offered a pro-natalist approach to the issue of childbearing. Noting the widespread fear of "race suicide" due to low birthrates among whites, he issued "an appeal to patriotism" to increase the size of allegedly good families. "Not only patriotism, but religion—our duty to God and man—also makes its appeal for larger families," he exulted. Not only God and country, but also allegiance to "the race" motivated Shannon to voice an imperative for the propagation of "larger families where both parents are physically, morally, intellectually, financially and hereditarily fitted for parenthood."[27] In hindsight however, wherever the rhetoric of the "fit" came

---

24. In Laughlin, *Eugenical Sterilization*, 324–25; cf. discussion in Kevles, *In the Name of Eugenics*, 92–95.

25. Rosen, *Preaching Eugenics*, 71–73.

26. Ellis, *Little Essays of Love and Virtue*, 102.

27. Shannon, *Nature's Secrets Revealed*, 204.

into play during the Progressive Era, the spectre of the "unfit" always lurked in the shadows nearby.

Conclusion

Today's Disability Studies literature allows a fresh window on the discussion of marriage restriction that had impetus within the eugenics movement a century ago. The widespread use of amniocentesis and the newer test "MaterniT21" so as to pressure pregnant women to abort their children based on the possibility of Down Syndrome is only one of many ways expertise is used to interfere with the sacred trust of life.[28] The rise of the "wrongful birth" lawsuit is another troubling echo of eugenic ideology persisting today.[29] The right to marry and to start a family is a fundamental right, as acknowledged in Article Sixteen of the universal human rights recognized by the United Nations shortly after this era.[30] Encroachments on that right were essential to the success of the eugenics movement.

In a 1931 message entitled "Light on Modern Marriage," Lutheran Theologian Walter A. Maier criticized the inadequacies of social scientific solutions to marital difficulties. "We do not believe that uniform divorce laws, stricter marriage regulations, vacations from married life, courses in eugenics, trial marriages, blood tests, and similar suggestions will lead to the desired results," Maier proclaimed.[31] Such advice remains a sage reminder that traditional marriage has long been under siege and is each generation in need of articulate public defenses by those who seek out ancient wisdom for families struggling amidst a sea of moral confusion.

Adherents of degeneration theory, including scientists and ministers, thought that social control over marriages and over family formation could be exerted for the eugenic improvement of American society. Moral and physical fitness for marriage were not the only considerations held dear by social reformers. Another key malady that motivated many social reformers to oppose freedom of marriage among the poor was the oft-cited "menace of the feeble-minded." To an exploration of that fraught concept our narrative now turns.

28. Leach, "The Reckless, Profitable Elimination of Down Syndrome."
29. See Lysaught, " Wrongful Life," 9–11.
30. "The Universal Declaration of Human Rights," Article 16.
31. Maier, *The Lutheran Hour*, 274.

CHAPTER 5

The Degenerate Mind and Hereditary Mental Defect

The condition of human beings, reduced to the extremist states of degradation and misery, cannot be exhibited in softened language, or adorn a polished page.

—Dorothea Dix, 1843[1]

Introduction

IN THE AMERICAN REVOLUTIONARY and post-revolutionary eras, care for the cognitively disabled fell mostly upon the families of persons thus afflicted. Historian Allison C. Carey has traced the development of concepts of rights for the intellectually challenged in U.S. history. The application of rights language to such persons was a very slow historical development, and not a story of linear progress. During the early nineteenth century, a rudimentary concept of rights for the mentally disabled gradually emerged. One way rights for the disabled came about was through the law's recognition of a distinction between competent and incompetent persons under the umbrella of those of "slow wit." Some persons deemed slow but essentially competent could retain such rights as voting (if male), marriage and procreation, and ownership of property. Others deemed incompetent could be restricted on all these activities. Still, it was often difficult for persons with disabilities to initiate a legal claim to their rights, and a relative or representative of sound mind was required to start such procedures on such a person's behalf.

---

1. Dix, "'I Tell What I have Seen,'" 622.

Institutional options for those who had varying levels of incompetence were rather limited. Still, for those who had no family able to provide care, nineteenth-century almshouses and hospitals began to house a wide range of dependents including the feebleminded. Almshouses gained acceptance more quickly in urban areas than in the rural. Still, the stigma was strongest on those regarded as able-minded or able-bodied enough to work but unwilling to do so. Such persons were often regarded as the "unworthy poor" in contrast with the "worthy poor," namely, those truly mentally incapable of gainful employment. Social reformers such as Dorothea Dix were by mid-century describing for the public horrible and cruel conditions that she witnessed in many institutions for the insane and the cognitively disabled. By the latter half of the nineteenth century, mental illness became a state concern, and larger institutions with increasingly professional and specialized staffs gradually became normative. Even amidst increased attention to the care of persons with mental disability, Carey notes that social norms continued to place women and minorities at a disadvantage in such institutions.[2]

By the late nineteenth century, reformers arose within the ranks of caregivers for the cognitively disabled. By moving away from the almshouse model and into a more professional and institutional mode of care, many believed that this environmental change could gradually improve the mental conditions of some to the extent that re-integration into society might occur. Beginning with the first specialized institution in Massachusetts in 1847, the emphasis was on education and training, with an accent on inculcating values of courtesy, industry, thrift, and self-discipline, and a weeding out of bad habits. Carey notes that the rise of eugenics thinking stressed "feeblemindedness" as a "biological and hereditary condition leading to the downfall of society." Thus by the turn of the new century, the emphasis was on keeping the feebleminded apart from the larger society, and away from opportunities to procreate.[3] This attitude toward the feebleminded, along with increasing efforts to categorize levels of mental ability in a scientific manner, had many practical outcomes for the treatment of such persons through much of the twentieth century.

2. Carey, *On the Margins of Citizenship*, 37–46. See the discussion of almshouses in the Northeast in Ferguson, *Abandoned to their Fate*, 21–44.

3. Carey, 49–51.

Hereditary Mental Defectives

In his 1898 address to the Conference on Charities and Correction, James C. Carson M.D. made an impassioned plea to keep the feebleminded from procreating. As Superintendent of the Syracuse New York State Institute for Feeble-minded Children, perhaps he had reached a level of frustration others could only begin to appreciate. Regarding feeblemindedness as a category of disease, indeed as an epidemic, the fear in his rhetoric is jarring. Citing a litany of statistics to underscore a sobering reality, Carson noted that the 1850 census had numbered the feebleminded at 681 per million. By 1890 that number was 1,526 per million. Leaving aside the historical likelihood of more exacting reporting mechanisms in 1890 than in 1850, the impact on Carson's mindset toward these unfortunates can only be described as one of horror. Estimating that in his own day only 7,000 of 100,000 feebleminded in the populace were under institutional care, Carson lamented that "the existence of so many feeble-minded establishes a centre from which emanates an almost endless chain of evil." Describing these disfavored denizens of the underclass as both a burden and a menace, he saw them as a fountain of additional forms of mental and moral weakness: "insanity, epilepsy, pauperism, illegitimacy, and every form of degeneracy."[4] As a prescription for this flow of unwanted individuals, Carson set forth the institutional solution in the clarion language of degeneration theory:

> With reference to the wide-spread army of degenerates scattered here and there throughout the land, and who are annually adding to the general plethora of feeble-mindedness, we believe that humanity, economy, the protection of society, and the prevention of degeneracy demand the permanent sequestration of the entire body of the feeble-minded within our borders in institutions. Especially should safe custody and State guardianship apply to every feeble-minded girl and woman of the child-bearing age. Besides these restraining influences mentioned, ought we not to expect great and growing benefits to future generations by the compulsory study and a thorough teaching of physiology, hygiene, and the laws of heredity in all of our higher schools and colleges?[5]

Such rhetoric proliferated during this time period. In 1903 patent attorney Casper Lavater Redfield (1853–1943) published *Control of Heredity: A Study of the Genesis of Evolution and Degeneracy*. Redfield took for

4. Carson, "The Prevention of Feeble-mindedness," 297.

5. Ibid., 303.

granted the division in society between classes. He gave voice to a fear pervasive among elites of his time:

> Common observation and the statistics of marriages, births and
> deaths tell us that the ignorant, the vicious, and the mentally
> incompetent individuals of a community marry early and rear
> large families, while the intelligent and desirable members of society marry late and have few offspring. The result of this is that
> the descendants of the ignorant classes are becoming relatively
> more numerous and threaten to supplant the descendants of the
> intelligent class.[6]

The assumptions embedded in this short statement were to play out on multiple fields. Positive eugenics was the idea that "intelligent and desirable members of society" ought to marry young and produce many children. President Theodore Roosevelt had six children, and urged similar efforts at fecundity upon old-stock Americans as a moral duty.[7] The threat of supplanting, however, could not merely be addressed by positive eugenics. Ways of restricting the undesirables from reproducing included marriage restrictions, segregation from society and from contact with fertile members of the opposite sex in mental institutions, and at times sterilization of those deemed "unfit" by elites.

Redfield devoted a chapter to expanding the definition of degeneracy. He defined the term more loosely than the theory's founder Bénédict Morel had, and thus applied the degeneracy descriptors to a wider swath of the population. "In man the word degeneracy is used to express any retrograde condition, such as deformity, or any congenital weakness of body or mind. Thus, idiots, insane and weak minded persons, epileptics, and the criminal and pauper classes are designated as degenerates."[8]

In the earlier investigations by Morel, degeneracy had more narrowly to do with what we today would call cognitive disability, not the broad catch-all phrase "any retrograde condition," of body or mind. Further, "weak minded persons" would not have been a helpful descriptor in Morel's studies due to its vagueness. Epileptics, paupers, and criminals were certainly problems in Morel's day, but none of them necessarily showed clear evidence of multi-generational physical or hereditary degeneracy. The evidences of degeneration Morel investigated spanned at least four generations, with ostensible specific and observable pathological deteriorations ending in sterility. Redfield's definition of degeneration held none of Morel's

---

6. Redfield, *Control of Heredity*, 64.

7. Morone, *Hellfire Nation*, 273–74.

8. Redfield, *Control*, 206.

cautious parameters. Thus, degeneration theory had by the turn of the century morphed in public policy discourse into a totalizing construct used to categorize greater and greater numbers of people as a mass of the unfit who should be made subject to elite social control.

The chapter went on to recount the Ishmaelite family (pseudonym for a dysgenic family investigated in Indiana by nineteenth-century eugenicist Oscar McCulloch), and the Jukes family (pseudonym for a dysgenic family investigated by Richard Dugdale in the 1870s in the same state). Redfield offered a unique interpretation based upon the notion that children born to older parents tend to be more advanced down the path of degeneration than children born of young parents. This rationale evinced the lingering effects of the doctrine of the inheritance of acquired characteristics in the early twentieth century. Under Redfield's theory, the longer an adult lives the more dysgenic characteristics and habits they accrue, when they sire children after accruing said characteristics, they blight their offspring with more bad habitual tendencies.

As noted in chapter 2, August Weismann had offered a cogent alternative to the inheritance of acquired characters over a decade earlier. Redfield had already discussed Weismann in a very cursory way, and had rejected the concept of the stability of the germ plasm.[9] Redfield averred that offspring should be born before parents reach 25, and "children produced after the age of 30 are more than likely to be tainted with degeneracy arising from parental viciousness."[10] Redfield's chapter on degeneration bore scant resemblance to the theory under that name originally promoted by Bénédict Morel. The protean nature of the terms "degenerate" and "degeneration" in this period is well illustrated in Casper Redfield's writings. Other experts brought their clinical and managerial experience of working with the cognitively disabled into the growing discourse of their eugenic control.

Unchecked Vicious Heredity

Alexander Johnson, Superintendent of the School for the Feebleminded in the state of Indiana chaired the Committee on Colonies for Segregation of Defectives at the 1903 Conference of Charities and Correction. Perhaps as evidence of its origins as a committee document, the text oscillated between harshness and compassion in its rhetoric toward the cognitively disabled. The document also ironically admitted the imprecision of its categories of analysis in one paragraph, while shortly thereafter asserting its scientific

9. Ibid., 56–58, 72–73.
10. Ibid., 214.

probity. Sounding a cautionary note, the document intimated that: "There are few even in this intelligent assembly, who do not exhibit some of the so-called stigmata of degeneracy." As a balm to mitigate this potential insult, the document hastened to affirm: "The degenerates whom we have in mind are those who, either physically, or morally, are so far below the normal that their presence in society is hurtful to their fellow citizens, or that their unhindered natural increase is a menace to the well-being of the state."[11] The document went on to define degenerates in the broadest possible manner, listing some 18 problems ranging from the chronic insane to tramps to the shiftless poor to "some deaf-mutes" to consumptives. Consciously recognizing that society could not segregate or control such a vast number of persons, the committee tried to focus on "certain classes of defectives who are either truly, hereditary degenerates or whose condition resembles this so much that they may be treated like them"; i.e., with "complete and permanent control."[12]

The committee admitted that when "asked for statistics of the degree of heredity among various classes of degenerates" that "trustworthy data are . . . difficult to collect." They added that even the census takers for the U.S. government in 1870, 1880, and 1890 had been unable to collect such data because "the attempt to collect them was offensive to relatives" along with others who had to be interviewed for such information. Even institutions in the private sector did not do an adequate job of collecting such data. Still, the committee accepted the figures from "careful people" working with the feeble-minded that heritability of this "most typical class of degenerates" was as low as 50 and as high as 95 percent.[13] Notwithstanding such useless, glaringly inconsistent statistical pabulum, the committee confidently wrote: "If the opinions above expressed are well founded, it follows that in some way or other the fatal heredity should be brought to an end." They added plaintively, "How can we possibly leave the world better for our work if we do not at least begin to stop this vicious stream at its fountain head?"[14]

The document went on to paint a rather rosy and idyllic picture of the goal of establishing rural colonies for the feeble-minded and other ill-defined categories of degenerates under the auspices of state sponsorship. Yet the committee could only point to three states as exemplary in taking such an initiative up to that point: Ohio, Wisconsin, and Massachusetts.[15]

11. Johnson, "The Segregation of Defectives, 245.

12. Ibid., 246.

13. Ibid., 247–48.

14. Ibid., 248.

15. Ibid., 250.

By contrast, the committee painted a dismal picture for existing alms-houses at county or town levels. The religious rhetoric is palpable: "Here may be found victims of the sins of their parents, of their own vices, of unchecked vicious heredity." Further, the cost of care for such persons, whether at the state level or through local or private charity, meant that "the degenerates live a parasitic life. They are, and must be, supported by their abler brethren." Yet despite this cost, the committee sermonized: "Some of us believe that while the teachings of the Carpenter of Nazareth have in them the most sublime, spiritual significance, yet their chief significance appears when they are applied to the common affairs of life." Further, "Some of us believe that the 'little ones' of whom Jesus spoke, are, for every man, those within his influence who are younger or weaker, or more sinful or more suffering than he." To fail to care for such persons would be worse than "that a mill stone be hanged about our neck and that we be cast into the depths of the sea, unless we have done all that man may do to abate the evil and the wrong, to protect the simple and the helpless."[16] Biblical allusions abound in the narrative, indicating that the Protestant ethos of biblicism continued to hold sway in the charities and corrections orbit.

The committee rejected sterilization, regarding it as a solution far into the future and dreadful to contemplate. In actuality, legislatures were then debating sterilization, and Indiana would become the first state to pass involuntary sterilization into law a mere four years later. The committee regarded sterilization as not in consonance with civilization, and a return to Roman or Spartan values rather than American values. "Our theories of life are inseparably bound up with the belief in the infinite value of the individual human being," they loftily intoned. They added, for good measure, the biblical trope: "Our weaker brother is of immensely greater value than the sparrow which falleth to the ground, or the beasts that perish." Finishing their critique of sterilization with another metaphor, the committee announced: "We may not do with him as we do with our cattle, for the benefit of ourselves or of the state."[17] Many eugenicists of that time strongly disagreed with such compassion, deeming it sentimental and counterproductive.

H.H. Goddard and the Sterilization Question

In his study of the oft-proscribed Kallikak family, H.H. Goddard gravely noted the problematic nature of trying to rein in the fecundity of the feebleminded. First he noted that average citizens often would not be able to

16. Ibid., 251–53. The mill stone reference is an allusion to Matthew 18:6.

17. Ibid., 249.

recognize the feeble-minded in their midst. This underscored the need for professionals like Goddard and his researchers to perform the task. Secondly, he noted the problem of caring for "this large army of people." The answer was for Goddard a combination of segregation into colonies of the feebleminded (again affirming the need for institutions such as his), coupled with better identification of those who do and do not need such care. While Goddard had earlier used very deterministic language to describe that defective branch of the progeny of Martin Kallikak, he admitted that "the hereditary factor" was operant in only 65 percent of his cases (while remarking that others held it to be 80 percent).[18]

In 1912 Goddard opposed sterilization as an answer to the breeding practices of the feebleminded, though he left the question of its future use open to further investigation. In some moments of candor Goddard admitted as a serious difficulty "the determining of which persons were proper subjects to be operated upon." In this light, he also confessed that "we are still ignorant of the exact laws of inheritance." He stipulated that the heritability of mental characteristics and the means of their transmission across generational lines "is not definitely known." These gaps in knowledge posed for Goddard obvious unease. "It therefore becomes a serious matter to decide beforehand that such and such a person who has mental defect would certainly transmit the same defect to his offspring," he cautioned, "and that consequently he ought not to be allowed to have offspring."[19]

In a concluding section, Goddard promised a future study designed to apply the insights of Mendelian genetics to the issue of feeble-mindedness. He speculated that perhaps feeble-mindedness did indeed follow straightforward Mendelian ratios, with feeble-mindedness as a recessive trait. Since this simplistic scenario (if true) would yield a ratio of 3 normal to 1 defective offspring in procreation, Goddard was hesitant to embrace sterilization as a method of redress. After all, such a regimen would deny life to 3 normal persons in order to prevent one abnormal individual, and this for H.H. Goddard in 1912 was too much to ask of society on the basis of a barely-understood science.[20]

Another family study, based on cognitively disabled families in Minnesota published some seven years later, cited Goddard's research. Mendelian principles provided the framework for the strong heritability of mental traits. This study assumed that "feeble-mindedness is a recessive trait due to the absence of a determiner for normality in the germ plasm." On this basis,

18. Goddard, *Kallikaks,* 106.

19. Ibid., 109.

20. Ibid., 109–16.

the study concluded that "feeble-minded parents would have only feeble-minded children." The marriage of a normal and a feeble-minded person would yield normal children, but all these "would be capable of transmitting feeble-mindedness to their offspring." Even if these normal offspring married normal mates, the Mendelian ratio of three normal (dominant) to one feeble-minded (recessive) would result on average. The study added the caution that the low number of human offspring made it statistically difficult to apply Mendelian ratios to human beings. It also admitted that "it seems improbable that so complicated a thing as general intelligence can be considered a unit character."[21] These qualifications did little to dampen the authors' enthusiasm for using heredity to explain in a simplistic manner the mental deficits of residents of the pseudonymous "Vale of Siddem."[22]

By 1923 Goddard was still reticent to see involuntary sterilization as a solution to feeblemindedness to be practiced on a wide scale. He preferred colonization and segregation by gender as the default mechanism for making sure that the feebleminded did not procreate. Goddard argued in favor of education and training, and opposed those who regarded the feebleminded as ineducable. To those who were enthused about sterilization, he stressed the environmental factors in their disability. "But again, says some one, they will always be vicious and dangerous and a menace to society," Goddard intoned. "There are, however, no facts to prove this," he countered, adding that many of feeble mind had been "misunderstood and mistreated."[23] Whereas many eugenicists had a special fear of the "high grade feeble-minded" due to their ability to blend into society and produce children of low intelligence, Goddard believed these to be the most hopeful cases. "May it not be possible," he pled, "that we will find use for all these people of moderate intelligence, and that the production of so many high grade feeble-minded is only the production of so many more people who are able and willing to do much of the drudgery of the world, which other people will not do [?]." Yet in a tone of resignation in the face of the popularity of sterilization, Goddard conceded "We must have sterilization wisely

21. Rogers and Merrill, *Dwellers in The Vale of Siddem*, 12–13.

22. This is a biblical reference to a very unpleasant locale near Sodom, full of tar pits, into which victims fell and perished, see Genesis 14:10. The implication is that feeble-mindedness is a hereditary quagmire from which rescue is nearly impossible.

23. Goddard, *Feeble-mindedness*, 587. In that same year University of Missouri Sociologist Charles A. Ellwood held that "those who are not normal in their hereditary endowments should refrain from marriage," and that the unmarried "should lead lives of continence." For others, Ellwood called for institutional restriction, namely, "all who are so abnormal that they cannot be controlled by the moral standards of society should be segregated in institutions and supported at public expense." Ellwood, *Reconstruction of Religion*, 200–201.

and carefully practiced for the solution of many individual problems that are not reached by any other method."[24]

Amos Butler and Indiana's Burden of the Feebleminded

Amos W. Butler (1860–1937) was a walking talking institution in social reform circles in Indiana and around the nation in the early decades of the twentieth century. Secretary of the Board of State Charities in Indiana, Butler would go on to serve as president of the National Conference of Charities and Correction in 1907. The September 1911 edition of *Indiana Bulletin of Charities and Correction* contains Butler's 1907 Presidential Address to the conference. His opening remarks showed a flair for the dramatic: "We bear the burdens of our civilization." If anything, this was a Progressive Era truism, and echoed countless speeches and sermons of the day concerned with social deterioration and its amelioration. Butler held the period in which he lived to be especially onerous compared with previous eras. "Our civilization, however, brings with it its burden of defectiveness and degeneracy," he warned, "our form of human society preserves and cares for the sickly, the weak, the dependent, the offender, and the defective. We have dependents of various kinds, of varying degrees, who do not fit into the family life, the school or the social system. They are unable to bear the struggle of life or the weight of responsibility that is imposed upon citizens. Their burden then becomes ours"[25]

The shouldering of such a burden was for Butler a duty both religious and scientific. He quoted the Apostle Paul to the effect, "Ye who are strong ought to bear the infirmities of the weak." Yet the overall tenor of Butler's remarks indicates that the process of bearing such burdens meant eliminating certain rights of persons identified as feeble-minded or otherwise burdensome on society. Butler's audience could be called citizens; while inhabitants of Indiana's various institutions for the less fortunate he held at a certain rhetorical and legal distance from the full rights of citizenship.

Butler focused most of his remarks upon the problem of the feebleminded. This raised the issue of definition that bedeviled numerous eugenicists during this period of history and drove the phenomenon of intelligence testing over the ensuing decades.[26] While such efforts to categorize and sub-categorize mental defectiveness were cloaked in scientific

24. Goddard, *Feeble-mindedness*, 588–89.

25. Butler, "President's Address," 297.

26. See Kevles, *In the Name of Eugenics*, 76–84, 129–31.

terms; to today's reader the terminology is at best quasi-scientific; at worst pseudo-scientific. "Ordinarily those included in the term 'feeble-minded' range from children who are too dull to learn in the public schools to the most helpless idiot." As if to add specificity to his analysis, Butler continued: "Feeble-mindedness, imbecility and idiocy are simply varying degrees of arrested mental development, all being classed as feeble-mindedness." This and similar befogging statements indicate that to social reformers like Butler, the burden of scientific accuracy yielded too often to the politics of oversimplification and even obfuscation. "Persons who are really feeble-minded are sometimes adjudged insane, and placed in a hospital for the insane," could seem to be "an injustice to the insane, for whom the accommodations thus occupied should be reserved." While the distinction Butler offered here was important, the speech contained no discussion of how said distinction was to be articulated.

Butler's rhetoric concerning the so-called feeble-minded then took a decidedly negative turn. He inveighed against "the blight of feeble-mindedness." "The debasing and demoralizing influence of an unrestrained feeble-minded woman in a community is beyond the comprehension of the uninformed," he observed. Many recent studies have pointed out the degree to which eugenicists focused their efforts on the control of women. The most famous Supreme Court case in eugenics history, Buck v. Bell, for example, would come to focus in 1927 on whether or not a young woman named Carrie Buck had a right to procreate.[27]

Even the married were not immune from criticism. Butler blasted the marital practices of the feeble-minded, especially the state expense of "the increasing population of imbeciles and defectives" that resulted from such unions. Dehumanizing rhetoric, seen elsewhere in eugenics literature, recurred in Butler's contention that "Each of these, if he lives, is destined to spread blight and the germs of disease not only among his kind, but among families of intelligence." Illustrating the way religious tropes were adaptable to eugenic ends, Butler recurred to religious language again, predicting that: "The results are increasing idiocy, epilepsy, physical infirmities, the stigmata of such defectiveness."[28]

The balance of Butler's address cited a litany of statistics indicating that the burden comprising the subject of the address was measurable in dollars and cents. Butler applauded the small number of states who had "established custodial institutions for the detention and care of feeble-minded

27. The most exhaustive treatment is Lombardo, *Three Generations, No Imbeciles,* 149–73.

28. Butler, "President's Address," 298.

women during the reproductive period." The discipline such institutions imposed had several salutary effects in Butler's opinion, namely that "they have regular habits, their strength is employed in useful service and their passions are restrained." Blunt, succinct, and jarring are the terms in which Butler matter-of-factly dispatched these ostensibly problematic women: "Forever they are separated from the world. Forever they are prevented from reproducing their kind."[29] Many were the nameless 'they' subjected to such segregation and isolation; but to Butler, this was a beneficial and wise course that he urged other states to adopt.

Among the solutions Butler enjoined to resolve or mitigate the burden of the mental defective were "correct marriage laws and means of preventing the reproduction of degenerates." He commended several states which had moved "toward more scientific measures," and which had passed "more advanced laws which tend to prevent the marriage of the unfit." Butler called for uniformity among the various states in updating their marriage restrictions.[30] He applauded not only custodial solutions to the reproduction of the feeble-minded, but also Indiana's involuntary sterilization law, passed in that same year of 1907.

Butler was a reformer thoroughly convinced that feeble-mindedness was a dire social menace, and the fountainhead of a host of social ills. "Feeble-mindedness produces more pauperism, degeneracy and crime than any other one force," he fulminated. "Its cost is beyond our comprehension," he concluded, adding, "It is a burden we are compelled to bear; therefore let us bear it intelligently, to the end that the chain of evil may be lessened, the weak cared for, and the future brighter with hope because of our efforts." He ended his address as it had begun, with a citation from the Bible: "Ye who are strong must bear the infirmities of the weak." Thus a profound cognitive dissonance between the rhetoric of blame and the rhetoric of compassion intermingled in Butler's speech. Such terminology was tragically commonplace at the dawn of the twentieth century.

On May 12, 1908, Amos W. Butler delivered an address before the National Conference of Charities and Correction, which met at Richmond, Virginia. His remarks filled the opening pages of the September 1908 issue of *Indiana Bulletin of Charities and Correction*. "The Board of State Charities and the People," was the title of his speech. Here he made programmatic statements about the Board of State Charities, of which at this time he was serving as secretary. Butler's rhetoric, like that of many eugenicists, swept a wide range of problematic persons together, and his words

29. Ibid., 302.
30. Ibid., 303.

vacillated between a cold harshness and a warm compassion. Concerns of "the people" indicated in the title of this address were likely in his view the taxpaying citizens of the middle class. Those for whom the state had to provide received more specific, and pejorative, labels. Butler wrote: "A Board of State Charities studies the whole charitable and correctional field. All the anti-social elements are but parts of one great subject. The insane, the prisoner and the pauper are but visible manifestations of human activity that may have sprung from similar causes or even the same cause. By studying underlying causes the Board could seek measures "to prevent pauperism, degeneracy and crime." [31]

To drive home his opinions, Butler wrote disparagingly of the family heritage of two sisters who had been inmates of a state institution for some 20 years, and whom Butler anticipated would be there until their deaths. He described them as "two of four illegitimate children born in the poor asylum to a degenerate woman, who was herself a public charge."

He lamented that other members of this family had been allowed to marry, "encouraged by the ease with which relief was obtained," and further castigated "an indifferent public" who had "left it possible for their children to grow up in the midst of ignorance and degradation indescribable." He further lamented that "the natural result was untold misery and shame and public expense." Butler lauded studies of such conditions that had resulted in various efforts at social reform. Such measures included: "stricter marriage laws; laws requiring better poor asylum management, improved poor relief administration, the segregation of feeble-minded women, the asexualization of confirmed criminals, idiots and rapists," as well as "state supervision of dependent children." Butler anticipated positive results from such measures for the future of the citizens of Indiana.[32]

Butler also spoke of the need for coordination between state and private/religious charity and corrections work. He urged that: "the churches form a natural way to reach the public. The state or district organizations of the different denominations can very properly each have its standing committee on charities. This is the practice in some states-why not in all?" With particular attention to prison conditions, Butler suggested that congregations hold an observance the last Sunday in October, called "Prison Sunday." He described the goals of such an event as follows: "That day is given to presenting prison conditions and the results of work with prisoners, the best ideas of prison reform, the causes of crime and the preventive

31. Butler, "Board of State Charities and the People," 2.

32. Ibid., 2–3.

work being done."[33] As a rationale for such an observance, Butler offered words both theological and pragmatic: "The churches could be interested in the local problems and their co-operation sought; for, after all, the practice of charity is the practice of religion. Man in his religious nature demands something vital in which to believe and a method of expressing that belief. What therefore will appeal to him more than rational methods in practical philanthropy? It is only necessary that he be shown the way."[34] Butler went on to recommend the involvement of other social organizations, such as colleges, teachers' associations, medical societies, The Bar Associations of states, and the State Conference of Charities, with their holding of local conferences on related topics.

Butler served as President of the Indiana State Conference of Charities and Correction in 1915. Butler's Presidential Address in Richmond, Indiana was entitled, "At the End of the Century," where he surveyed the 100-year history of charity and corrections in the state of Indiana. None were so qualified to assess the work of charity and correction than the redoubtable Amos Butler, whose voluminous correspondence with national and state-wide leaders of orphanages, homes for the feeble-minded, jails, and prisons indicate a keenly activist spirit. As a Quaker, Butler tapped into a rich tradition of activism, especially in prison reform.

That Butler was a proponent of a largely decentralized, local approach to the care of the unfortunate is borne out by his correspondence on behalf of the Indiana Board of Charities. Butler's letters reinforce this impression, as he carried on a continuous and aggressive letter-writing campaign to the county officials in charge of jails and other institutions. Further, his correspondence indicates continuous efforts to recruit volunteer citizens to serve on the local boards overseeing such relief work, including a patent willingness to enlist the clergy of various denominations into community leadership through such boards. Thus the relationship of church and state in the arena of care for the cognitively disabled was at this time in Indiana a profoundly interdependent one.

Nests of Social Incompetence

In 1911 A.C. Rogers, Superintendent of the Minnesota School for Feeble-minded and his Research Assistant, Maud A. Merrill, began a series of family studies in the hills of Minnesota. The subjects of the study were several generations of a few families who intermarried and lived in "a certain valley"

33. Ibid., 3.
34. Ibid., 4.

the authors dubbed "the Vale of Siddem." With chagrin the authors noted that "in one county we found such numbers of feeble-minded and degenerate people that we have selected that section for special investigation."[35] Rogers and Merrill used deprecatory rhetoric that was the common parlance of eugenic family studies. In focusing on the "Yak family," the authors intoned that "Their moral standards, their petty thievery, neglect of decency, carelessness and vindictiveness stamp them, even when they are not mentally deficient, as the gravest sort of social menace." The authors lamented that the standards of civilization were breaking down in the upper Midwest, even though their patriotism made it hard to admit:

> The commonwealth of Minnesota, young and vigorous, harbors already such nests of social incompetence, degenerates, defectives and criminals as existed in the Juke's ancestral mountain fastnesses. Mental deficiency is indigenous to the same soil that produces criminality, sex laxity, alcoholism and pauperism. Whatever the relation of cause and effect in the matter, the sociological evidence is indisputable.[36]

Conclusion

Degeneration theory had the power to weld together quite disparate social problems and give them a scientific/theoretical framework. In such a framework both scientific and religious language found a common cause where social reformers could seek solutions to issues like mental disability.

Hereditary mental defect was often constructed as a degenerative quality that brought multi-generational devastation to families and to society as a whole. Often reformers saw it as the ground of other social problems such as poverty, sexual promiscuity, and criminality. With the rise of large institutions charged with the care of the feebleminded came the effort to make them the subject of scientific study and control. Despite the presumed objective scientific basis of policies aimed at social control of the cognitively disabled, religious and moral rhetoric still permeated the literature.

The eventual decline of the popularity of eugenic explanations for cognitive disability cannot be explained definitively. Some possibilities may be suggested. The extremes of eugenics policies toward disabled persons in Germany became public beginning in the 1930s and 1940s, evoking international revulsion, though the extent of the practices was only fully realized

---

35. Rogers and Merrill, *Dwellers in The Vale of Siddem*, 9.

36. Ibid., 24–25.

later.[37] The leveling effects of the Great Depression may also have chastened the class bias of at least some elites who suddenly found themselves in the ranks of the middle class or the poor. Prominent psychiatrists began asserting a more compassionate outlook toward those with disabilities, and pharmacological interventions began to mitigate some symptoms of some mental disorders. As more scientists began to pull away from the eugenics agenda, its ability to continue the demonization of the feebleminded was concomitantly curtailed. Still, attitudinal change toward cognitive disability was, and is, a slow-moving process in American culture.[38] This is illustrated in the next chapter as we explore the treatment and maltreatment of those afflicted with epilepsy.

37. See the harrowing description of Nazi Germany's T-4 program for the liquidation of handicapped persons in Evans, *Third Reich at War*, 75–108.

38. *Special Olympics International*, "Rosa's Law."

CHAPTER 6

Epilepsy and Eugenics in Scientific
and Religious Perspective

*There is no class that needs more sympathy and the kindness of the
community more than the epileptics. This is a disease about which
the medical profession knows almost nothing. There is no disease
of which the medical profession acknowledges itself more ignorant.*

—Rev. H.H. Hart, 1903

Introduction

SOCIAL REFORMER MARY JACKSON Ruffner set before the West Virginia legislature, on more than one occasion, plans for an "Asylum for Incurables" in
the late 1890s. The institution was approved for construction in Huntington,
West Virginia by the end of the century. Her contemporary, epilepsy expert
William Pryor Letchworth, while in favor of such institutions, nevertheless remonstrated that: "To commit deformed and crippled children or
epileptics, many of whom are improvable, if not curable, to a place having
so disheartening a designation as Asylum for Incurables, would be a sad
mistake."[1] Letchworth's comment in 1900 bespoke the optimism of a new
century and the social reform ethos of Progressive Era professionals. Yet
epilepsy has continued to pose challenges to families up to the present day.
The literature of the eugenics movement routinely lumped persons with
epilepsy in with cognitively disabled persons with whom they could be dramatically dissimilar. The concept of "the unfit" was beginning to take root in
social reform circles, and could serve as a leveling term of opprobrium that

---

1. Letchworth, *Care and Treatment of Epileptics*, 180.

neglected to treat the individual patient as an individual.[2] Epilepsy became one of many symptoms of "degeneration" to be brought under the control of American eugenicists.

What to do with epileptics? The question haunts the cultural narrative of epilepsy, as well as theories of its causes and cures. Stigmatization, segregation, and eventually sterilization all had their discriminatory impact on persons with epilepsy in the late-nineteenth and early twentieth-century setting. Separation from larger society was often the solution envisioned and favored by experts, but the exact forms and dimensions of the institutions developed to facilitate such segregation varied widely. Depersonalization was a sad side effect of the increasingly large scale, inherently bureaucratized institutionalization of the treatment of epilepsy in the late nineteenth and early twentieth century.[3] Efforts at scientific study and efficiency in treatment of epileptics surely had their merits, but also their deficits, in the amelioration of this often poorly understood condition.

Epilepsy is defined as "a recurrent and paroxysmal disorder starting suddenly and ceasing spontaneously due to occasional excessive rapid and local discharge of the nerve cells in the grey matter (cortex) of the brain."[4] Determining the cause of the phenomena of seizures remains a daunting task, for in a large majority of cases no cause is definitively identified, yielding the term "idiopathic epilepsy." Seizures are manifested in a host of ways, including "episodic impairment or loss of consciousness, abnormal motor phenomena, psychic or sensory disturbances, or perturbation of the autonomic nervous system." Sometimes localized or widespread lesions of the brain can be blamed.[5]

In the reform literature describing late-nineteenth and early-twentieth century efforts to improve the treatment of epileptics, a wide number of treatments were prescribed. Often social reformers classed epileptics with other ostensibly problematic persons who required institutionalization. By the 1870s many physicians and asylum managers were recognizing the

2. For a book-length treatment of the term, see Carlson, *The Unfit*, 39–94.

3. On the negative impacts of institutionalization on residents, gaining notice as a formal problem in social science by the late 1930s, see Johnson and Rhodes, "Institutionalization," 219–36.

4. "Epilepsy," *Black's Medical Dictionary* 4th ed., 243.

5. "Epilepsy," *Dorland's Illustrated Medical Dictionary* 29th ed., 607. See also *The Dictionary of Psychology*, 335, for entries on the following symptoms: epileptic absence, epileptic aura, epileptic character/personality, epileptic clouded states, epileptic cry, epileptic dementia, epileptic deterioration, epileptic equivalent, epileptic focus, epileptic furor, and epileptic stupor. Under "epileptic character/personality" the reader finds a minority of patients described as "irritable, stubborn, egocentric, uncooperative, and aggressive."

inadequacy of housing epileptics with the insane or with the "feebleminded." Thus began a slow and scattered effort in a few states to tailor remedies to the predicament of epileptics in a more precise fashion.

Various proposals for the treatment of epileptics were set forth in the *Proceedings of the National Conferences of Charities and Correction* from 1874 to 1912, as well as books and articles on epilepsy from the same period. The *Proceedings* begin in 1874, and the 1912 date marks the beginning of the use of Phenobarbital and other barbiturates to address epilepsy, in a decade marking a more effective chemical turn in the treatment of the disorder.[6] With the rise of professionalization and specialization came something of a shift over time from an explicitly religious ethos of care to a more clinical, secular, institutional ethos of social control. Attempts at specialized treatment for epileptics included sporadic attempts to distinguish epileptics from the insane and feebleminded, their segregation into large colonies, and arguments for marriage restriction and even sterilization prior to the rise of widespread pharmacological intervention.

Care for Epileptics Prior to 1900

According to medical historian Walter J. Friedlander, English neurologist John Hughlings Jackson was the father of the modern study of epilepsy. His definition in 1873 marked an important moment in the identification of the disease: "[Epilepsy is] a sudden, excessive, and rapid discharge of grey matter of some part of the brain." In 1876 Jackson clarified the nature of the disease as recurrent and chronic.[7] Because seizure phenomena can accompany several different maladies temporarily, and because not all seizures manifest in the same manner, by the 1880s clinicians could speak of *epilepsies*, rather than presenting it as one definite disease.[8]

The variety of proposed causes of epilepsy during the period in question made policies for the care of epileptics uneven and disjointed. For a time hyperemia, or increased blood flow to the brain, was debated as alternately the cause or the remedy for seizures. Mirroring this debate was the possibility that brain anemia, or decreased blood flow was the culprit. By the early 1890s such explanations began to lose favor. Similarly, theories about metabolic effects of nutrition on epilepsy were tried and eventually

6. "Phenobarbital," *Dorland's Illustrated Medical Dictionary*, 1369, defines this drug as: "a long-acting barbiturate, used as a sedative, hypnotic, and anticonvulsant, administered orally."

7. Friedlander, *History of Modern Epilepsy*, 2.

8. Ibid., 3.

discarded, with the notable exception of a few benefited by a ketogenic diet.[9] By the 1920s tests of cerebrospinal fluid began to point in the direction of what would later be called neurotransmitters and their role in the production of seizures.[10]

More promising for a time were investigations of patient reactions to external stimuli as a cause of seizures. A combination of sensorimotor factors played a role in the theory that epilepsy was a dysfunction of the nervous system. An 1870 study described the cause as "some unhealthy, and therefore irritating condition, acting upon the excito-motory system of nerves."[11] This theory thus had an impact on treatments that put much emphasis on creating a peaceful environment for epileptic patients, with opportunities to be outdoors, to be housed with those with like severity of symptoms, and to live in relatively small cottages. Such features of care were perhaps more an ideal than a reality within the spectrum of epileptic intervention, but did reflect experts' concerns with environmental factors in patient improvement.

The location of the origin of the seizures in the brain received intensive study during our period. Over time, consensus shifted from the medulla oblongata region of the brain (1860s–1890s) to a focus on the cerebral cortex as a more promising center of investigation by the second decade of the twentieth century.[12] The stage was set for neurochemical interventions designed to ameliorate the source of the seizures, the erratic activities of neuronal structures of the brain.

The neuron doctrine was articulated in a series of papers in Germany in 1891 by Wilhelm von Waldeyer, Director of the Anatomical Institute at the University of Berlin.[13] Identification of neurons as key to the communication of stimuli throughout the central nervous system was a breakthrough for the study of many disorders, including epilepsy. Over time debates over whether the function was electrical or chemical found resolution in the term electro-chemical. The possibilities of pharmacological intervention for epileptic seizures rose as chemical inhibitors to electrical impulses in the body were discovered through experimentation. The problem was, and still remains one of targeting. Chemicals produce side effects, sometimes quite negative, beyond the therapeutic effects sought. The harrowing history of the use of chemical restraints in twentieth century mental health is a story

9. Reuber, et al., *Epilepsy Explained*, 224.

10. Ibid., 14–18.

11. Ibid., 20.

12. Ibid., 21–23.

13. Ibid., 30.

for another time, but has become better known in recent years. We now turn our attention to efforts to care for epileptics in the period prior to its clear recognition as a neuro-chemical disorder of the cerebrum.

In 1881, Ira Russell, M.D. read a paper entitled "Care for Epileptics" at the National Conference of Charities and Correction (hereafter NCCC). Russell lamented the conditions epileptics endured and the inadequacy of their care: "Many are confined in poorhouses, some with maniacal complications are in insane asylums, others are cared for by friends at home, a few only receiving the care and treatment common humanity requires and the importance of the disease demands."[14]

Given the limited resources then devoted to the care of epileptics, coupled with a dearth of solid scientific information on their condition, Russell criticized the medical establishment for neglecting them. "Epileptics have been generally left by the regular medical profession," he fumed, "as the legitimate prey for quacks and charlatans."[15] In the essay Russell averred that almshouses and insane asylums, where many epileptics were being placed, were not appropriate venues for their treatment. He perceived the plight of epileptics as untenable, fulminating that: "It is deplorable when we contemplate how little interest has been taken in, and how little has been done for this class of human beings when so much has been done for the insane, the deaf, the blind, the inebriate, the idiotic, and feeble-minded."[16]

In the ensuing debate following the remarks of Dr. Russell, Dr. T.W. Fisher agreed that "separate hospitals for certain classes of the insane are likely to come into favor" when their numbers in given states reached a certain point. He further opined that "Superintendents have always found it difficult to classify certain classes of the insane, as, for instance, idiots and imbeciles, the criminal insane, epileptics and dipsomaniacs." He noted, further, that "epileptics are usually sane between the attacks, at which time they often resent companionship with the insane, while they are in turn most unpleasant and dangerous neighbors during the seizures."[17]

Further comments by General Brinkerhoff of Ohio indicated a growing consensus on the issue: "I think all our Superintendents are in favor of separate asylums, because they feel that epilepsy is a disease of its own kind, and that by having separate epileptic institutions there can be a better structure, a better classification, and better care."[18] These remarks were pre-

14. Russell, "The Care of Epileptics," 325.

15. Ibid.

16. Ibid.

17. Ibid., 326–27.

18. Ibid., 330.

scient, as the state of Ohio, spurred by the Board of State Charities, passed an act on April 11, 1890, resulting in the establishment of the Gallipolis Epileptic Hospital. The cornerstone was laid November 12, 1891, and the name established in 1892 as "The Ohio Hospital for Epileptics." This was the first such state-supported institution in the USA devoted solely to the care of epileptic clients. By April of 1912, the institution held 1475 patients on the colonization plan.[19]

In 1886, the NCCC heard a paper by Dr. George Knight entitled "The State's Duty toward Epileptics." Here we find an adumbration of what would come to be called negative eugenics, or the stopping of the procreation of epileptics. Knight's watchword was "prevention." The eugenic mindset constructed prevention of future "victims" as an act of positive compassion. Following is an example of the rhetoric typical of the time:

> One of the first and most important measures we must advocate would be such a framing of the laws governing marriage in our several States as would make the marriage of an epileptic a crime. From my own knowledge, I can cite the case of an epileptic woman who became the mother of fifteen defective children. Eight died in infancy from lack of vitality, two inherited epilepsy, two were fairly teachable imbeciles, and the other three had sufficient intelligence to marry and reproduce, according to the laws of heredity . . . . This instance is only another proof of what we must all believed, —namely that like produces like; and, therefore, as a natural sequence of things, epileptics in the majority of cases *must* produce defective children.[20]

The initial method of achieving prevention of the marriage of epileptics was known as segregation—a series of separate homes for epileptic males and epileptic females. Knight summarized his program with three terms: prevention, mercy, and economy. It is not difficult to discern which of these three aims would receive short shrift under such a policy.

So far the condition of epilepsy, either explicitly or implicitly, often got classed as a subset of insanity. This confusion persisted for some time, at least until the varieties of mental health disability could be analyzed by a growing cadre of specialists and professionals devoted to specified disorders. In 1890, Richard Gundry, M.D., repeatedly used the moniker "epileptic insane" as a distortive term in his essay "On the Care and Treatment of the Insane." "Of the epileptic insane I desire to say a few words. They are disturbing elements in every asylum," he warned. Emphasizing the sensational nature

19. *Asylum Projects*, "Gallipolis Epileptic Hospital."
20. Knight, "The State's Duty toward Epileptics," 299.

of some seizure events, he cautioned his audience: "However mild their forms of insanity generally, however amiable their character, they are liable to explosive paroxysms of fury; and their epileptic attacks are shocking to witness." He boldly added statistical weight to the problem by claiming "It is calculated that about one-third of epileptics are insane." He lamented that those who are not insane, "or are only momentarily so," tend to lose their employment, and get "driven from pillar to post," and finally "sink into a state of indifference and pauperism."[21] For a solution to this dismal state of affairs, Gundry drew his audience's attention to Europe, where 24 years prior to this a German Lutheran clergyman, Pastor von Bodelschwingh, established a farm colony for epileptics near Bielefeld, Westphalia. From 1867 to 1887, some 2,407 epileptic patients had received treatment at this colony. Medicinal intervention was part of the treatment, such that by 1890, "eight hundred pounds of potassic bromide are now consumed per month by residents and correspondents" (i.e., patients around the world who request such medications by mail).[22]

The facility divided patients into small "family" units in cottages, as well as the separation of the sexes, and segregation also by "age, social standing, former occupation, etc." Gundry noted that some of the housing units had been planned, bricks made, masonry laid, and wood-work and iron-work, painting and glazing had all been done by epileptic residents. Implicitly this demonstrated that epilepsy was not a mere form of insanity, and confirmed a growing consciousness internationally that specialized care was needed. Indeed some rather prominent geniuses were known to have been epileptics, including the novelist Fyodor Dostoevsky, the playwright Moliere, and the industrialist and peace-prize namesake Alfred Nobel.[23]

Returning to the American scene, Gundry was encouraged by the new work in Ohio for epileptics, as well as the work of Rev. Frederick H. Wines of Illinois in the establishment of a facility with 1,609 residents in Kankakee with organization akin to that of a village.[24] The religious fervor that marked much of the rhetoric of social reform in the Conference of Charities and Correction in the late nineteenth century emerged in Gundry's closing remarks. Taking the form of a benediction, they contrast with the conference's later sanitized and secularized language of bureaucracy:

> It is to be hoped that before the next meeting of this Conference steps will be taken in some portion of the land to found

21. Gundry, "On the Care and Treatment of the Insane," 263.

22. Ibid., 265–66.

23. Reuber, et al., *Epilepsy Explained*, 6.

24. Gundry, "On the Care and Treatment of the Insane," 267.

an institution of the character required, and that this will lead to the establishment of other institutions in every State in the Union. Such a work would advance our common humanity, and be crowned with the blessing of God.[25]

In 1896, William Pryor Letchworth (1823–1910) gave the address "The Care of Epileptics" at the NCCC. The address surveyed work among epileptics in several states of the union, highlighting those who followed "the praiseworthy example of some European countries in making special provision for epileptics."[26] Letchworth quoted from Dr. Henly Chapman Rutter, manager of the newly-established Hospital for Epileptics in Gallipolis, Ohio. Rutter boasted the establishment of a "nucleus for a large colony." Facilities included hospital care, "accommodations for imbeciles, idiots, and other helpless classes," with "places for worship and amusement for the entire community" as well as an administration facility. In line with the scientific research necessary to pursue a cure for epilepsy, he noted: "We have a well-equipped laboratory, and are supplied with scientific investigators, whose time and attention are devoted entirely to observations of the malady." He estimated recoveries at between 5 and 10 percent, while noting some uncertainty of success, "since we do not consider a patient recovered until at least two years have elapsed since the last seizure."[27]

Letchworth noted the work of the Passavant Memorial Homes at Rochester, Pennsylvania some thirty miles outside Pittsburgh, begun in 1895. He reported that "its immediate management is confided to an order of Deaconesses of the Lutheran church," and had as its model the German Lutheran experiment in epileptic care at Bielefeld.[28]

In describing the new facility for epileptics at Lapeer, Michigan, Letchworth implicitly acknowledged the specialization of care envisioned for feeble-minded and epileptic patients. While the money was not to be available until 1897, "the home, when completed, will care for epileptics separate and apart from the feeble-minded." Noting the rise of specialization consciousness on the part of some states, he projected that "it will be so planned as to effect a complete classification of the inmates in different buildings. Insane epileptics are at present cared for in the State asylums for the insane."[29]

Mary J. Dunlap, Superintendent of the New Jersey Institution for Feeble-minded Women spoke at the NCCC in 1899 on the theme "Progress

25. Ibid., 278.
26. Letchworth, "The Care of Epileptics," 199.
27. Ibid., 200.
28. Ibid., 202.
29. Ibid., 203.

in the Care of the Feebleminded and Epileptics." With evident pride, after surveying the work for epileptics in several other states, Dunlap boasted: "In New Jersey, my own State, the dream of an epileptic village has been realized; and during the coming year two cottages are to be erected, and the foundation and plans for a large institution carefully laid."[30] Important to the separation of epileptics from the "feebleminded" was research indicating the reality of genius along with epilepsy. She cited the research of Professor Hausemann of the University of Berlin, who had investigated the brain of mathematics genius Helmholtz, who "though epileptic and hydrocephalic" had been pronounced "the world's greatest mathematician, physiologist, physician, and natural philosopher during the past four decades."[31] Yet the full impact of the research had not fully dawned on the author, as her concluding remarks included once again the lumping together of disparate mental disorders under the banner of degeneration theory: "Our campaign is against the evils of degeneration that follow the neglect of idiots and imbeciles and epileptics. They cannot plead their own cause. Then how much heavier is our responsibility!"[32] Notwithstanding the great achievements of epileptics, still a pervasive paternalistic/maternalistic attitude toward them permeated the consciousness of many elites who exercised considerable, and increasing, bureaucratic social control over their lives.

Care for Epileptics 1900–1912

William Letchworth expanded his investigations of the care of epileptics into a book-length treatise, published in 1900. He used the language of degeneration theory to describe this persistent malady. Regarding heredity he warned that "this source of epilepsy, idiocy, and crime has reached threatening proportions." He added: "The vast number of degenerate persons that now burden society, whose defective organization is traceable to hereditary causes, and who continue to transmit their weaknesses—physical, mental, and moral—presents a difficult and important problem to social scientists."[33] In order to inhibit the procreation of insane or feeble-minded persons (with whom epileptics were routinely lumped), Letchworth applauded a Pennsylvania statute that penalized "any clergyman or civil officer who shall knowingly solemnize such marriage" with a misdemeanor conviction resulting in

---

30. Dunlap, "Progress in the Care of the Feebleminded and Epileptics," 257.

31. Ibid., 258.

32. Ibid., 259.

33. Letchworth, *Care and Treatment of Epileptics*, 9.

6 months in prison and a $500 fine.[34] Another prominent author on epilepsy at the time held views on this topic even more strongly worded. William P. Spratling, M.D. argued in 1904 that "marriage confers a license for the creation of a diseased progeny generally lower in mental, moral, and physical stamina than their antecedents. This fact alone should be sufficient to deny the epileptic the right of marriage in fully ninety-five out of every hundred cases in which it is sought."[35]

The classification problem recurred throughout Letchworth's tome. He lamented the lack of provision for epileptic children, "many of them bright and intelligent" who nonetheless ended up as public charges "in forced companionship with the lowest class of idiots."[36] He cited a host of experts in epileptic care calling for separate provision of care for epileptics, "having borne emphatic testimony to the necessity of providing homelike institutional care, with expert medical treatment, for this numerous and neglected class."[37] Such experts touted the wisdom of the colonization of epileptic patients due to their danger to public safety, and to bring under restriction "the procreation of degenerate offspring."[38] Letchworth believed that colonies should open with only a few patients and to have numbers increased only gradually. Indeed, experts routinely lamented overcrowding and large waiting lists for epileptic facilities, as a perusal of the *Proceedings of the National Conference of Charities and Correction* shows. Letchworth also urged discrimination between those who have a reasonable prospect of cure or improvement and those who are already beyond cure.[39] The balance of the treatise was given over to the investigation of work among epileptics in 18 states, plus England and three epileptic institutions on the European continent.

The lengthiest survey, including 24 pages of text and four pages of photographs, was The Ohio Hospital for Epileptics, at its founding in 1891 labeled the "Asylum for Epileptics and Epileptic Insane." Located on a sprawling property on the Ohio-West Virginia border in Gallipolis, Ohio, this facility captured Letchworth's sustained attention. The buildings depicted, at 3 stories and of imposing pressed-brick or buff-colored sandstone construction, appear larger than the small-scale cottage idealistically envisioned by the author. Separate men's and women's dining halls each held

34. Ibid., 13.

35. Spratling, *Epilepsy and its Treatment*, 302–303.

36. Letchworth, *Care*, 19.

37. Ibid., 20.

38. Ibid., 21.

39. Ibid., 30.

a seating capacity for 800 patients.[40] From its opening in 1893 to 1898, the hospital admitted 1,295 patients (some of whom were re-admissions).[41] The challenge of having sufficient staff to permit personalized and compassionate care in such a context can hardly be minimized in retrospect.

Letchworth acknowledged, albeit briefly, some problems encountered with local community residents. For example, he noted that "sewage is discharged into a tributary of the Ohio, but against the protests of residents and local boards of health." This led to appropriation of funds to "introduce an elaborate sewerage system" upon the plan of a similar facility in Massachusetts.[42] Diet of the residents was regulated carefully, in part because institutional scientists were exploring the potential role of diet in the etiology of epileptic seizures.

Though a state-sponsored facility, a rather open attitude toward religious activity characterized the facility in the late nineteenth century. Letchworth approvingly noted that "The prayer-meetings in the different houses and the regular religious services in the chapel, the Sabbath-school, and the entertainments in the amusement hall have been potent factors for good." This was designed to give patients "opportunities for social, moral, and religious improvement" like those of average citizens.[43]

Pastors from the city were allowed to conduct services in the facility, including Catholic mass. The influence of religion however was not always idyllic. The director candidly indicated that religious services would run smoothly for several months, then "there may be a demonstration of some kind." Letchworth recorded one incident thus: "At one time a discussion between a Baptist and a Methodist ended in a fight." For such reasons, "controversies are forbidden," though apparently they were not entirely avoidable.[44]

The role of religion in an increasingly professionalized context of care, such as a state-sponsored institution, became an increasingly delicate matter to maintain. As the field of charities and corrections became less dominated by clergy and its ranks filled with bureaucratic and scientific experts, the role for religious expression, for good or ill, became marginalized.

It turns out that Letchworth's account of The Ohio Hospital was somewhat sanitized compared with the actual reality. The Ohio General Assembly investigated allegations of patient abuse at the facility, the results of

40. Ibid., 73.
41. Ibid., 85.
42. Ibid., 78.
43. Ibid., 81.
44. Ibid., 83.

which were published in 1897. Further, a dispute arose in 1901 between the institution's chief scientist, Dr. A. P. Ohlmacher, and the Director, Dr. H.C. Rutter. This resulted in the governor's very public dismissal of Rutter under a cloud of charges of incompetence and concerns over the inflated salaries of Rutter and his family.[45]

At the 1901 National Conference of Charities and Correction, William A. Polglase M.D. summarized progress in the care of those afflicted with feeble-mindedness and epilepsy over the previous century. Polglase served as Medical Superintendent for the Michigan Home for the Feeble-minded and Epileptic in Lapeer, Michigan. His address, "The Evolution of the Care of the Feeble-minded and Epileptic in the past Century" noted the connections between Christian ideals of charity and improvements in care for such patients. Yet even at the beginning of the nineteenth century he lamented the treatment of patients in such institutions, including chainings, cagings, floggings, and starvation. By contrast he lauded recent progress, in "the science that now mitigates and cures 20 percent of cases of insanity," and noted "the humanity and love that have established schools and homes for the feeble-minded and colonies and hospitals for the epileptics" as products of the previous century.[46]

By 1901 he could boast that "In the United States to-day there are more than a score of large institutions for the care and training of the feeble-minded, and nearly a half-dozen colonies for epileptics; and provision for the establishment of several more is already made." Polglase complained that the sheer numbers of feeble-minded and epileptic citizens were rendering the size of such institutions inadequate, noting also a warning from science: "During the past twenty years scientific study of the defective classes has demonstrated that large numbers of criminals, inebriates, and other persistent offenders against law and morals are really congenital imbeciles." While for a time professionals had hoped for placement of many such persons back into society as productive members, they now realized the need for "lifelong detention of all defectives," and in almost all the newer institutions it had assumed the status of a fixed policy.[47] Due to economics, therefore, the colony plan had become normative.

Among epileptics, some 20 percent were feeble-minded according to Polglase. He bemoaned their lot in life, stating "the majority of so-called normal epileptic persons, because of poverty or ignorance, cannot or do not take the proper care of their health, and live careless or criminal lives."

45. "Gallipolis Epileptic Hospital."
46. Polglase, "The Evolution of the Care," 187.
47. Ibid., 188.

For such persons "the pitiable condition and unhappy state" as well as "their menace to the social welfare of the community" led to provision for their "special care." Here Polglase pointed with approval to the colony at Bielefeld in Germany. He applauded widening efforts to specialize epileptic care, highlighting "the first annual meeting of the national association for the study of epilepsy and the care and treatment of epileptics" gathered that very same week in Washington, DC.[48]

Attitudes of other social elites toward epileptics however lacked qualities such as compassion or comprehension. Francis Galton, father of the term *eugenics*, in a 1908 tome on human mental abilities placed "epileptics and other persons of instable [sic], emotional temperament" among the "criminal classes."[49] Because epilepsy was seen as a primarily hereditary problem, it became caught up in the rhetoric of the eugenics movement and agitation for social control of procreation, and in policies designed to impede epileptics (and others) from marrying and having children.

More even-handed in his assessment of the problem of epilepsy, leading epileptologist physician William Spratling (1863–1915) gave the treatment of epileptics a more fulsome analysis in his massive and influential 1904 tome *Epilepsy and its Treatment*. Here he subdivided the topic into three categories, devoting a chapter to each in turn: general treatment, medical treatment, and surgical treatment. His chapter on the general treatment of epileptics focused on colony life as "the final and most satisfactory plan of securing full control of the patient." He noted as other studies had the successes at the Lutheran institution at Bielefeld in Germany, as well as colonies in New York state and in London, England. The advantages of such an institution were

1. to provide a home life "simple and elemental in form."

2. to preserve individuality and

3. to provide vocations for those able to perform tasks ranging from simple to complex.[50]

Spratling's ideal for the size of such colonies was rather utopian at one acre per resident. Compassion appeared in Spratling's calls for freedom of movement, and the minimization of use of barred windows and physical restraints. He also recommended the colonies strive to achieve a strong level of independence from the broader community. His ideal colony included an

48. Ibid., 189–90.

49. Francis Galton, *Inquiries into Human Faculty*, 45.

50. Spratling, *Epilepsy and its Treatment*, 339–40. The book runs to 522 pages.

administrative building, a hospital, industrial buildings to include manufac-ture of the basic goods needed by the residents, and educational buildings. For women he recommended "laundering, sewing, knitting by hand and machine, darning, rug-weaving, basket-making, and the domestic arts," reflecting the gender sensibilities of the era. He also called for a laboratory for scientific research.[51]

In the chapter on medical treatment of epilepsy, Spratling made this frank admission: "As yet we are ignorant of any means of suddenly curing the disease," and expressed pessimism toward the likelihood of a cure. He further noted that: "unfortunately the influence of drugs is transient making necessary their repeated renewal." The chapter surveyed the use of bromine preparations, hypochlorization, codein, borax, nitroglycerin, chloretone, zinc, sodium biborate, urethane, solanum carolinense, tar derivatives, iron, chloroform, and serotherapy, in addition to the occasional usage of electricity and hydrotherapy.[52] The breadth of this list illustrates the highly experimental nature of the era's medical efforts to control epileptic seizures. In 1907 the leading British expert in epilepsy, Aldren Turner, found some efficacy for the use of bromides if employed as an early intervention at the onset of epileptic symptoms, particularly in young patients. This was the most popular medicinal intervention in the early decades of the twentieth century, having been used widely in epileptic care since the late 1850s.[53] But by the early twentieth century enthusiasm for it had begun to cool, due to toxicity, overuse, and various negative side-effects from doses too high.[54]

Surgery for epileptics was effective only in a limited number of cases of two specific forms (*grand mal* and Jacksonian epilepsy). Spratling offered brief case studies of the results of 34 surgical interventions from the Craig Colony in Sonyea, New York. In 27 of the cases cited the result was either no permanent improvement or, in a few instances, an actual worsening of the epileptic condition.[55]

According to a retrospective issue in the prestigious journal *Epilepsia*, the first decade of the twentieth century experienced a dearth of original

---

51. Ibid., 344–45.

52. Ibid., 359, 362.

53. "Bromide salts (mainly potassium bromide) were the first effective antiepi-leptic agents found . . . With the advent of newer anticonvulsants, and recognition of their toxicity, bromides had largely faded from use by the middle of the [twentieth] century. In recent years there has been some interest in resurrecting them for patients whose epilepsies prove resistant to all other appropriate therapies." Eadie and Vajda, eds., *Antiepileptic Drugs*, 212.

54. Shorvon, "Drug treatment of epilepsy," 70, see 73.

55. Spratling, 404–11.

or ground-breaking research in epilepsy. Between 1907 and 1910 the British journal *Brain* had no articles primarily on epilepsy; the major related French-language journals had only 3 original articles, and the major German journal concerned with psychiatry and neurological disorders held only 4 articles on epilepsy. To fill this void, the journal of the International League Against Epilepsy was founded in 1909 in Budapast. Its journal *Epilepsia* was first published under the auspices of the ILAE in 1909.[56] This was about the time that states in the USA began legalizing the sterilization of the allegedly "unfit," a category into which epileptics were routinely cast by enthusiasts of the eugenics movement. [57]

Charles B. Davenport, America's premier promoter of eugenics in this era, authored a 1911 study entitled *A First Study of Inheritance in Epilepsy*. As noted above, Davenport was the son of a Congregationalist minister, but eventually touted eugenics as a new religion. In keeping with Davenport's more general favoring of hard hereditarian determinism, it is not surprising that his study of epileptics in several epileptic institutions yielded the following conclusions, among others. Davenport concluded that the behaviors of the feeble-minded and the epileptic alike supported "the hypothesis that each is due to the absence of a protoplasmic factor that determines complete nervous development." He further urged that "when both parents are either epileptic or feeble-minded all their offspring are so likewise." He warned that if marriage among epileptics continued at its current pace with "no additional restraint" then "the proportion of epileptics in New Jersey would double every thirty years." He therefore concluded that "the most effective mode of preventing the increase of epileptics that society would probably countenance is the segregation during the reproductive period of all epileptics."[58]

Davenport's right-hand man for running the eugenics empire known as the Eugenics Record Office was Harry H. Laughlin (1880–1943). As superintendent of the ERO, Laughlin engaged in a relentless effort to make eugenics central to the policies of the USA regarding "the feebleminded." His leadership of the ERO, from 1910 to 1939 can hardly be overestimated in terms of impact on American social policy. He authored a model involuntary sterilization law whose language was adopted by many states. His role as a eugenics expert was pivotal in the institutional identification of Virginian Carrie Buck as a "low grade moron" and "a potential parent of

---

56. Shorvon, "Early History," 1–2.

57. See e.g., Smith, "Marriage, Sterilization and Commitment Laws," 364–70; Strode, "Sterilization of Defectives," 296–301.

58. Davenport and Weeks, *First Study of Inheritance in Epilepsy*, 29–30.

socially inadequate or defective offspring."[59] Her right to procreate was eventually denied by the U.S. Supreme Court in *Buck v. Bell*, which opened the door to involuntary sterilization nationwide after 1927. Laughlin and Supreme Court Justice Oliver Wendell Holmes, Jr. agreed on the wisdom of involuntary sterilization. It was Holmes who authored the 1927 decision, including the following sentence: "It is better for all the world, if instead of waiting to execute degenerate offspring for crime, or to let them starve for their imbecility, society can prevent those who are manifestly unfit from continuing their kind."[60]

The son of a minister in the Christian church (Disciples of Christ) fellowship, Laughlin did not take to the fundamentalist doctrines of his father, according to historian Daniel Kevles. Laughlin's primary interest was science, and applying the wisdom of science, especially eugenics, to society. His eugenics ideology led him, in 1920, to work fervently to draft a sweeping immigration restriction bill. The eventual law, *The Johnson Immigration Act* of 1924 was built largely on the scientific racist principles derived from Laughlin's research.[61]

Historian Edwin Black notes that Laughlin's Eugenics Record Office scrutinized epilepsy as a "menace" because researchers routinely conflated feeblemindedness and epilepsy as hereditary, and thus intractable, maladies. Laughlin disseminated to many hospitals his own thirty-page tract, with ostensibly scientific charts and graphs, concluding that epileptics had to be stopped from procreating. Yet the definition of epilepsy employed by Laughlin and the ERO was highly imprecise and failed to account for the role of head trauma in many cases, stressing the hereditary factors instead.

Tellingly, late in life workers at the ERO reportedly witnessed Laughlin having seizures. He and his wife never had children, even though under eugenics ideology they should have produced many children due to their intelligence and allegedly superior stock. This leads Black to theorize that Laughlin secretly suffered from epilepsy. Black highlights the irony of Laughlin's opposition to epileptics, noting "Among his many crusades, Laughlin may best be remembered for his antagonism toward epileptics." His tragic and hypocritical demands for their segregation and sterilization only serves to add poignancy to the story of the mistreatment of those suffering from epilepsy.[62]

59. Lombardo, *Three Generations*, 134–35. The centrality of Laughlin to the involuntary sterilization and immigration restriction movements is succinctly surveyed in Cohen, *Imbeciles*, 103–59.

60. Ibid., 165–68.

61. Kevles, *In the Name of Eugenics*, 102–104.

62. Black, *War on the Weak*, 55, 395; cf. Reilly, "Laughlin, Harry Hamilton,"

Conclusion

This chapter was inspired in part by my viewing at a regional film festival of the short documentary *Last Pyramid*, about a mother's response to the death of her epileptic son at the tender age of sixteen in the early twenty first century. Epilepsy still has no cure, but the need for individualized care, compassion, and sensitivity highlights the plight of families struggling with a disorder often perceived as beyond cure.[63] Care for persons with epilepsy as well as the research underlying medical intervention has progressed since the period investigated here. Historical perspective indicates that religion and science have cooperated and can cooperate in holistic care of persons with a chronic disorder that afflicts 1 percent of the population.[64] Reminding ourselves of the need for positive regard for the dignity of patients as persons needing care as well as cure is a vital starting point. In her historical study of the treatment of the mentally ill by Protestant leaders, theologian Heather Vacek reminds us:

> Stigma clouds theological wisdom and misshapes Christian reaction to mental illness. However, awareness of the reach of stigma's tentacles brings conflicts between belief and practice to light, allowing a reconsideration of social stigma and theological claims that helps undergird a reshaping of Christian practice.[65]

The treatment and care of persons with epilepsy is only one dimension of the intersection of the history of the eugenics movement and institutional histories in the American context. The next chapter addresses the intersection of degeneration theory, eugenics, and the institutional context surrounding the rise and fall of the notion of the hereditary criminal.

American National Biography (1999), 13:252–53.

63. "Kevin's Pyramids," http://www.kevinspyramids.com/the-film.html.

64. Reuber, *Epilepsy Explained*, 4.

65. Vacek, *Madness*, 164.

CHAPTER 7

From Sinful to Criminal: The Making of Hereditary Criminality

The type of reformatory prison to be required, therefore, is less that of a reform school for juvenile, venial, and accidental derelicts, and more a scientific training place for degenerate adults.

—Zebulon Reed Brockway, 1912[1]

Introduction

SPECIAL ATTENTION IS GIVEN in this chapter to degeneration theory and the interplay of scientific and religious terminology utilized in identifying and controlling criminals by Protestant elites. The hereditary criminal has long held a dreadful sway in the minds of the leaders of society. No less a figure than Francis Galton, coiner of the term eugenics, confidently asserted that: "It is . . . easy to show that the criminal nature tends to be inherited."[2] Galton cited the famous study of the dysgenic Jukes family in America, published in 1876, noting with horror the fecundity of the Jukes, resulting in "no less than 540 individuals of Jukes blood, of whom a frightful number degenerated into criminality, pauperism, or disease."[3] Lumping together very disparate social problems with diseases was a typical *modus operandi* for Galton as for most eugenicists, illustrated by the fact that he placed "epileptics and other persons of instable [sic], emotional temperament" among the "criminal classes."[4]

1. Brockway, *Fifty Years of Prison Service*, 374.
2. Galton, *Inquiries*, 43.
3. Ibid., 44.
4. Ibid., 45.

The original Jukes family study of the Reverend Richard Dugdale, however, did not claim a clear hereditary or biological taint of criminality. Tragically most references in eugenics literature to the Jukes study claim it as proof of hereditary criminality. The Jukes study became if anything a stock argument in the eugenics arsenal, making the hereditary criminal a virtual truism. Nearly a century later the *Concise Dictionary of American Biography* of 1964 still asserted that social economist Richard Dugdale, author of the Jukes study, favored hard hereditarianism. The brief essay concludes, erroneously, that "Dugdale believed inheritance was more important in determining character than environment."[5]

While Dugdale acknowledged that heredity played a role in select criminal cases and some mental aspects of criminality, his emphasis was on environmental factors that contributed to pervasive criminality over several generations of Jukes. Dugdale concluded that his study showed "that heredity depends upon the permanence of the environment, and that a change in the environment may produce an entire change in the career, which, in the course of greater or less length of time, according to varying circumstances will produce an actual change in the character of the individual." Dugdale went on to criticize harshly the penal system which, in his words, "masses an army of moral cripples, cursed with contaminating characteristics held in common." He opposed the practice of congregating such persons in close proximity, which, in his view, "prepares an environment of criminal example just fitted for the assimilative power of each individual malefactor."[6] Overcrowded prison conditions did more than heredity to school the imprisoned in the ways of habitual crime. Eugenics historian Edwin Black accurately notes that "His book was quickly hailed as proof of a hereditary defect that spawned excessive criminality and poverty—even though this was the opposite of what he wrote."[7]

It is true that Dugdale's rhetoric was aligned with another feature common in his era's assessments of criminality: the use of moralistic and religious language. For instance, he accentuated the role of fornication among the Jukes as "backbone of their habits." He also freely referred to women in the Jukes lineage as guilty of "harlotry," which he regarded as a lesser variation of imprudent sexual misconduct compared with "the professional debauchee" whom he designated by the term "prostitute."[8] When behaviors of those on the margins of society became not merely sinful but

---

5. *Concise Dictionary of American Biography*, s.v. "Dugdale, Richard Louis."

6. Dugdale, *Jukes*, 113–14.

7. Black, *War Against the Weak*, 24.

8. Dugdale, *Jukes*, 13, 18.

criminal, efforts intensified to curtail said behaviors. The social control of such problematic persons increasingly became the purview of leaders in the criminal justice system, figures such as Judge Sanford Green.

Judging Crime as Sin

In 1889 Michigan Supreme Court Jurist Sanford Green (1807–1901) published a detailed overview of crime. In it he echoed the moralistic religious rhetoric common to such literature. He pronounced that: "The treatment and cure of those moral diseases from which all sin and crime proceed are subjects of more interest and importance to mankind than all others. Sin may never be entirely eliminated from this world, but I believe that, by means which have been proven to be practicable, it may be reduced to a small fraction of its present magnitude. . . ."[9]

Throughout his tome, Green applied biblical lessons seamlessly with his social policy analysis. Green, a progressive in the field of criminology, did not believe that heredity was "a tyrant from which there is no escape." Rather, he held that "the discovery of an existing inherited taint of disease or vice in a child" could be a boon for society. Such a discovery, if made early enough, could lead to an awareness of "lines of physical and moral treatment which may lead to the practical enfeeblement of the taint or even to its eradication."[10] Such a blending of hard heredity with the anodyne of social reform was a persistent incongruity in the arguments of many eugenics enthusiasts.

Social reformers working in the areas of urban health reform, penology, law enforcement, and the judiciary increasingly warned the public about the dire consequences of the "born criminal." New experiments in penology marked the end of the nineteenth century. One historian has described it thus: "The emergence of positivistic criminology and the medical model enhanced the prestige and power of superintendents and staffs; it opened a new scientific discourse in the field of criminal justice; it sanitized human and social engineering."[11]

9. Green, *Crime*, 6.

10. Ibid., 346.

11. Pisciotta, *Benevolent Repression*, 27.

Religion and Early Criminology

Many individuals trained for the Protestant ministry pursued the scientific impulse by entering the social sciences. In part this was a response to the burgeoning social problems that attended urbanization in the early twentieth century. Deep concern over the social problems of the city got impressed upon young ministers by such prominent Social Gospel promoters as Washington Gladden. The social sciences were situated adjacent to the professional ministry, given the parallel concerns of ministers and social sciences about the problems of urbanization in America. The National Conference of Charities and Correction brought ministers together with social scientists, penologists, physicians, and social workers. Some obviously ministerial roles, such as prison chaplaincy, bridged the professional gap. Key works in the secondary literature acknowledge the special affinity of criminology with Protestant theological modes of thought.

Nicole Hahn Rafter identifies the eight most influential authors endorsing the view that criminals are "physically distinct, atavistic human beings" during the critical years in the professionalization of criminology, 1890–1910. She notes that two of those eight authors, August Drähms (1849–1927)[12] and Charles R. Henderson (1848–1915),[13] were ordained ministers. A third, Arthur MacDonald (1856–1936), was theologically trained as well.[14] Drähms built upon a distinction in early criminology between two schools: the older, namely, "the Spiritualistic or classic" school, and the newer, "Positivist, or materialistic" school.[15] Over time, these two impulses in criminology remained, but the "spiritual" became more and more deeply implicit, and the "positivist" impulse became the predominantly explicit ethos, much like any number of other disciplines that grew out of the theological soil of the nineteenth century.

Historian William Burger traces the lingering impact of Puritan thought upon the historical development of American criminology. He

---

12. Drähms was resident chaplain at the State Prison at San Quentin, California. Drähms, *Criminal*, frontispiece.

13. Henderson was a Baptist minister and a sociologist. He served as a pastor in Terre Haute, Indiana and Detroit, Michigan. He entered the academic life in the dual role of chaplain and professor of sociology at the University of Chicago, serving there from 1892–1915. Among the more influential books he authored were *Introduction to the Study of Dependent, Defective and Delinquent Classes* (1893); *Modern Methods of Charity* (1904); and *Social Duties from a Christian Point of View: A Textbook for the Study of Social Problems* (1900). See *Concise Dictionary of American Biography*, s.v. "Henderson, Charles Richmond."

14. Rafter, *Creating Born Criminals*, 116–17.

15. Ibid., 111.

concludes that: "the history of American criminological thought can best be understood as an attempt to rationalize and secularize certain key elements of Puritanism."[16] In Burger's estimation, the Puritan effort to create a purified America, through an application of theological categories of inclusion and exclusion (covenant, election, double predestination, the realized kingdom of God on earth, the jeremiad against human pride and greed, etc.), has continued to hold sway throughout the history of criminology. The insider/outsider dichotomy, however, has increasingly been secularized and rationalized, and the kingdom-building work has increasingly come under the aegis of the state. Burger's analysis indicts these trends, referring to the effort of the first generation of sociologists "to unite sociology, church and state" as an "unholy alliance." Burger opines that Charles R. Henderson, who, as noted above, was a minister who became a leading figure in first-generation criminology, "viewed crime as evidence that America's Christian morality had been destroyed." Henderson's solution was to modify capitalism so that greed and exploitation were displaced by the second great commandment, namely, love of neighbor. By the second generation of criminologists, the emphasis was less on theological categories and more on scientific solutions to the problem of crime.[17]

Rafter's and Burger's analyses expose the need to understand more fully the theological ideologies at work in the constructions of the etiology of criminology among Protestant social reformers in the late nineteenth and early twentieth centuries. They also point to an important figure firmly embedded in the European antecedents of American criminology.

Theorizing the Born Criminal:
Lombroso Fathers a Field

The father of the field of criminology, Italian physician and penologist Cesare Lombroso (1835–1909), influenced greatly the efforts of social reformers to explain the problem of criminality. His book *Criminal Man* went through five editions in just twenty years (1876, 1878, 1884, 1889, 1896–97). Lombroso expressed some skepticism about the importance of degeneration theory in his third (1884) edition. Lombroso was concerned that its proponents were overreaching by attributing even "insignificant symptoms of illness" to degeneration. Still, he admitted that an "overwhelming number of psychiatrists" were embracing the explanatory power of degeneration.[18]

16. Burger, *American Crime and Punishment*, 1.

17. Ibid., 176–77.

18. Lombroso, *Criminal Man*, 221. For a detailed contextualization of Lombroso's

By the 1896 edition Lombroso would explain criminality as a "degenerative disease," and cite as evidence the Juke family, the infamous family wrongly alleged by eugenicists to be hereditary criminals. Lombroso wrote of the Jukes that their "great fecundity was followed by high infant mortality and finally by complete sterility, as also happens after the coupling of monsters or members of distant species."[19]

Later, in a manner that seems quite incoherent with the previous claim, Lombroso argued that "criminals share with geniuses a type of degeneration that produces not only evil but also new virtues." This was put to good use, Lombroso argued, when criminals showed the keen ability to "see more clearly than average men the defects of current governments."[20] How this had anything to do with "degeneration" is unclear, but the statement illustrates the protean explanatory power that degeneration theory had taken on by the end of the century.

Robert A. Nye has noted the dependency of Lombroso on the previous work of degeneration theorist Bénédict Morel. He writes: "Indeed, Lombroso's theory depended heavily on references to degeneration theory to account for the appearance of atavistic symptoms such as feeble-mindedness and moral idiocy that were characteristic of born criminals." Nye points out that by the late nineteenth century French criminologists were engaged in a fierce debate over heredity and environment, nature and milieu, and free will in discussions of the causes of criminal behavior.[21]

Degeneration theory was powerfully influential, especially among criminological theorists like Lombroso, and the pragmatic social reformers who applied his insights. A key bibliography of literature on social reform and crime, published by the U.S. Government in 1893, is replete with French, German, and Italian titles reflecting the prevalence in the social sciences of degeneration theory.[22] As the popularity of eugenics began to wane, in part due to the criticisms of geneticists as well as cultural anthropologists and religious (mostly Roman Catholic) critics, so did the popularity of degeneration theory by the 1930s. Below are a few examples of the interplay of religious and scientific themes represented by degeneration thought. Such themes found an especially fertile soil in the literature of this period that pertained to the growing belief in inherited criminality.

---

work and the debates in which he engaged, see Pick, *Faces of Degeneration*, 109–52.

19. Ibid., 327; cf. Black, *War on the Weak*, 24–25; Kevles, *In the Name of Eugenics*, 93.

20. Lombroso, *Criminal Man*, 352.

21. Nye, "Heredity or Milieu," 350.

22. See MacDonald, *Abnormal man*, 207–434. See also Clark, *Prevention of Racial Deterioration and Degeneracy*, 3–17.

Crime as a Phase of Sin

The Salvation Army was an evangelical organization dedicated to caring for the urban poor. The urban context often suffered from various levels of crime, and the intersection of criminological and theological investigations of human nature is evident in the Salvation Army literature. In 1901, the *War Cry*, the main organ of the Salvation Army, ran a multiple part series of articles entitled "Sin: The Hydra-Headed Monster" over several successive issues of the magazine. The series analyzed sin in "phases," or successive levels of degeneration that the author believed constituted a downward spiral ever lower into more egregious forms of sin. The fourth installment was entitled "The Crime Phase." The author cited with approval the researches of the Italian criminologist Cesare Lombroso, the promoter of the "doctrine of criminal heredity, or of innate propensity or pre-disposition to crime." So authoritative were Lombroso's findings, in the view of this editorialist, he could confidently claim that in the medical world the "'criminal head' is now a recognized and definite quantity in the study of criminals." In actuality, phrenological approaches to explaining criminality were on the wane. Phrenologists and craniologists believed they could discern intellectual acumen from the contours of human skulls, sorted by greater and lesser intellectual capacities as inferred by the size of the brain cavity. By 1900 biologists were abandoning this theory, but some reformers continued to assert its scientific probity and social utility.[23]

The author further averred: "Certainly it would appear to be a fact ascertained beyond power of contradiction that crime in the men and women of today is, at least often, appreciably the result of pre-natal influence." The negative ramifications of human habits, particularly in the form of alcohol abuse, were that civilization itself was threatened by hereditary criminality. The author lamented: "A hideous arraignment, this, of the world, after nineteen centuries of our boasted Christian civilization." He concurred with Lombroso's call for governmental intervention to remove the possibility that such persons could "be begotten, born, and reared in an atmosphere of crime." Here both the hereditary and environmental causes of criminality

---

23. On the rise of Phrenology and its influence among eugenicists, see Hilts, "Obeying the Laws of Hereditary Descent," 62–77; and Fabian, *Skull Collectors*, 24–26, 89–97. Hilts observes of the phrenologists: "They shared with eugenicists the conviction that the key to human progress lies in the improvement of man's biological endowment by attention to breeding, and they feared, as did the eugenicists, the possibility of racial degeneration" (73). See also the classic rebuttal of phrenology, Gould, *Mismeasure of Man*, 19–175; 351–64. Historian of Science Frank N. Egerton avers that "The last serious attempt to defend phrenology was W. Mathieu Williams's *A Vindication of Phrenology* (1894)"; see Egerton, "Phrenology and Physiognomy," 640.

were explicitly conjoined, but the physiological component, particularly the size and shape of children's heads, loomed large. The author voiced his agreement with Lombroso. "It is impossible to doubt that he is right," he declared, "for if crime be indeed a disease it can be cured only by absolute extirpation." Yet the author later cited the work of the Salvation Army in the San Quentin, California penitentiary as evidence that the gospel can bring about dramatic change in the life of criminals. The problem of the "phase of sin" known as hereditary criminality had contained a heavy physiological component in the minds of scientific criminology experts such as Lombroso. But for this evangelical editorialist, the solution to the problem of crime was the same as the solution to the problem of sin: the Christian gospel as proclaimed by the Salvation Army.[24]

Stigmata and Sterilization

In the first two decades of the twentieth century prison reformers began to reshape penology by the application of "eugenic prison science." The language of congenital defectiveness began to be applied to greater numbers of inmates. Sterilization as a solution to the "hereditary criminal" became more commonplace as a proposed solution.[25] A 1910 paper by Julius B. Ransom, M.D., physician to Clinton Prison in New York, encapsulated well the growing fear of the fecundity of criminals among reformers. Ransom's occasional use of religious terminology to render the threat of the hereditary criminal all the more dire is not atypical of the prison reform literature of the Progressive Era. Based upon the assumption that criminal behavior is primarily inherited, Ransom and many other criminologists, prison physicians, and penologists urged the legalization of involuntary sterilization.[26] Ransom summarized seven major arguments used by advocates of sterilization by 1910, some three years after Indiana became the first state to legalize the procedure under the leadership of sterilization enthusiasts John N. Hurty and Harry A. Sharp.[27] In brief, Ransom's arguments for sterilizing criminals were:

24. Barker, "Sin," 5.

25. See Pisciotta, *Benevolent Repression*, 112–18.

26. Largent, *Breeding Contempt*, 1. By the early 1960s some 63,000 forced sterilizations had taken place in the U.S., and this is not to count many sterilizations performed illegally. For an enthusiastic and lengthy endorsement of sterilization by a criminologist of this period, see Mosby, *Causes and Cures of Crime*, 82–129.

27. Indeed, fact is very often more fascinating than fiction. The names are real. See Black, *War against the Weak*, 63–66.

1. The fact of the great number of public charges recruited from the defective classes.

2. That defects physical and mental are transmitted to the offspring.

3. That, if a defective marries a defective, the children will inherit the stigmata of both parents, and be of a more defective type still. The natural tendency is for the abnormal to mate with the abnormal, consequently defectives are rapidly increasing in numbers as well as becoming more pronounced in type.

4. That a large number of this class do not respond to moral or intellectual influences . . . [and despite knowledge of the wrong] still lack will power to resist the impulse to commit wrongful actions.

5. That this class of persons is prolific, knowing "no law of self-restraint." These appear to be incapable of admitting their lack of ability to care for their offspring. . . .

6. That the restriction of propagation is necessary for the relief of this condition, and marriage restriction laws have proven inadequate.

7. That the absolute segregation into colonies and institutions of so great a number of existing "defectives" would require spending enormous sums of money.[28]

Two aspects of this list bear closer analysis. First, Ransom made the commonplace assertion and sweeping generalization that both physical and mental defects are transmitted to the offspring, resulting in criminal families. He did not qualify this proposition by what sorts of physical or mental defects are hereditary, or what would constitute a defect. Further, he included no statement of probability of transmission: he assumed the transmission to be absolute and without exception. Such totalizing language was necessary to justify involuntary sterilization. To admit that some offspring of criminals could turn out to be decent citizens would be to undermine the ostensible scientific basis of sterilization.

Secondly, Ransom completely turned the concept of "stigmata" on its head. Stigmata were, after all, the signs of Christ's crucifixion which many generations of pious Catholics believed could be seen anew in the bodies of especially holy saints, such as Saint Francis of Assisi. The term stigmata as used here became symbolic of extreme evil. Such usage subverted a religious term and changed its meaning to the complete opposite of its original meaning. Thus a word had been denuded of its pious content and turned into a propaganda tool for ruthlessness toward both criminals and their

28. Ransom, "The Prison Physician and His Work," 286–87.

children. As noted in chapter 3, the term "stigmata of degeneration" became a standard and oft-repeated trope in the eugenics literature. Stigmatization as a cultural meme thus has manifested over time shifting shades of meaning, being transformed from a religious to a quasi-scientific connotation, to eventually bearing a strongly prejudicial semantic load.

Regeneration as New Heredity

Albert H. Currier was a pastor in Oberlin, Ohio, once a hotbed of evangelical social reform fervor. In a 1911 article, he cited statistics ostensibly showing a dramatic increase of the number of inmates in state prisons, from one for every 1647 of the population in the year 1860, to one in 720 by the year 1900. Noting hyperbolically that America led the world in crime, Currier enunciated several causes for this problem. Included in his analysis were such causes as "unrestricted foreign immigration," "lack of industrial education and training," and "traffic in strong drink," among many other putative sources of the crime wave. Currier's assessment was not entirely gloomy, however, and he cited some felicitous trends (in his view) toward prison reform, such as the indeterminate sentence and the creation of a separate juvenile system. Citing an address by a Dr. Henry Hopkins before the National Prison Association in Kansas City in 1901, Currier identified religion as "the crowning reformatory agency." Noting that crime flows from two chief causes, heredity and environment, Currier quoted Hopkins approvingly to the effect that Christianity addresses both. "The man born from above, the regenerated man, has a new heredity," Hopkins had adverted. "The man who is 'in Christ Jesus' has a new environment." Currier pointed to the conversion of well-known criminals at the time to bolster his claims for the efficacy of the Christian gospel as central to addressing the problem of crime.[29] This was more *nurture* than *nature* and thus indicated the enduring claims of social and religious reform even in an increasingly scientific field.

Many social reformers were struggling to forge an adequate explanation for what they perceived to be an epidemic in crime in America. Protestant ministers, theologians, and editorialists interpreted crime through their theological categories. Yet a theological approach to hereditary criminality did not guarantee univocity among Protestant intellectuals and reformers. Some stressed the hereditary nature of criminality as an egregious instance of original sin. Others stressed the imperative of social reform in terms derived from evangelical perfectionism. Evangelical theological reflection

29. Currier, "Crime in the United States," 61–93.

provided categories, such as hereditary sin and social perfectibility that were particularly susceptible of distortion when viewed through the lens of eugenic ideologies.

An Unrelenting Tyranny of Ancestral Defect

Some in the penal field were less sanguine about religion as a partner in the treatment of those whose criminal tendencies were perceived to be inherited. Paul E. Bowers, chief physician at the Indiana State Prison authored an article entitled "Recidivism" in 1914. In reviewing the history of the treatment of the mentally ill, Bowers downplayed the sources of criminal behavior once promoted by religious institutions. "We no longer regard the habitual convict a demon-possessed unfortunate, or the wilful and conscious chooser of evil," Bowers began. Though once insanity was attributed to "the results of original sin," he went further and criticized the theological notion of crime as "deliberate sin." He noted that science in the modern world has "pushed through the crust of orthodoxy and delved into the study of those forces which regulate" the activities of criminals, and has declared "that he is mentally and physically defective." This defectiveness gives evidence of "pathological conditions due to defects of cerebral development or to acquired retrograde changes of the central nervous system."[30]

Such developmental or acquired changes surely begged for an analysis of the environment of the criminal, but Bowers' analysis would go on to place heavy stress on hereditary defect. He went on to evaluate the religious professions of prisoners in three categories:

1. Those who profess religion to impress prison personnel and hasten their release;

2. Those who use religion and its rituals as a kind of narcotic; and

3. "Those who earnestly seek, as far as their warped judgments and emotional instabilities will permit, a real reformation in character."[31]

Genuine conversion resulting from a rational assessment of the religion's truth-claims seemed not to have been a live possibility for Bowers.

Ironically Bowers rhetorically recurred often to religious tropes. For instance, he affirmed that the prison doctor is an "eternal proselyter" for reform and progress. In his discussion of "psychopaths," his analysis sounded very much like a variation on the older original sin theme, urging that:

30. Bowers, "The Recidivist," 404–405.

31. Ibid., 414.

"these unfortunate moral defectives we generally find to be burdened with an evil heredity, a harsh, unrelenting tyranny of ancestral defect." He further went on to castigate "sexual perverts of the most disgusting type" among the psychopaths in his prison, in language reminiscent of countless sermons on sexual sin. [32]

Indeed, Bowers's statistical analysis of 100 Indiana prisoners, if accurate, bespoke a condition worthy of pessimism. Sorting these subjects into categories that were loose, overlapping and hardly scientifically rigorous even for that date, Bowers left no prisoner uncategorized. The prisoners were grouped as follows: 12 insane, 23 feebleminded, 38 constitutionally inferior, 17 psychopaths, and 10 epileptics. The largest and most ill-defined category, "constitutionally inferior," stood above the feebleminded group, but only barely. He described this group as "below par both mentally and physically." The characteristics of the constitutional inferiors represented a laundry list of attitudes and behaviors: indecision, inability, vacillation, dependence, indulgence in prostitution, weakness for alcohol and drugs, mentally slow, while "his reason and judgment are defective," having "an unsatisfied craving for continual and unusual excitement," leading such a person to "live on the borderline of insanity and criminality." Strongly influenced by wanderlust, "they travel from place to place and the railroad employees and detectives often catch them breaking into boxcars to secure food and small plunder." The treatment Bowers suggested included removal of the physical results of "dissipation and venereal disease," taking them out of the company of "vicious and bad associates," and re-education for societal usefulness. By doing away with unproductive activity, and jobs "that would in a measure afford them novelty and excitement, their rehabilitation might be expected if it were not for their neuropathic organizations."[33] The degree to which many of these same characteristics might also describe rather normal 18 to 35-year-old males in the wider populace, analogous to a control group, did not garner any attention in the article.

Given the power of hereditary defect, Bowers, although giving lip-service to treatment, undermined that cause with a rather bleak determinism. Among his several negative conclusions, the harshest of them embraced degeneration theory root and branch. "Because it has been shown by eminent authorities that heredity is the greatest factor in the production of insanity, epilepsy, feeblemindedness and other neuropathic states," Bowers warned, "and because these conditions on the whole, when of the degenerative type, respond but little to treatment, the reformation of the chronic offender is a

32. Ibid., 405, 411, 412.
33. Ibid., 410–11.

high-sounding illusion."[34] He went on to insist that such persons should be studied and sorted into categories, and put in the charge of psychiatrists to determine their treatment. That his fatalistic analysis had already enervated any serious motivation for treating such prisoners escaped his attention.

Historian of Criminology Nicole Hahn Rafter observes the importance of degeneration theory to the growth of the notion of hereditary criminality in Europe and America at the end of the nineteenth and beginning of the twentieth century. She points out that "degeneracy was pictured as a tendency to devolve to a lower, simpler, less civilized state." Theorists who pondered the problem of degeneracy in the waning years of the Victorian era believed that "the downward spiral into degeneracy could be brought on by immorality (e.g., drinking, gluttony, or sexual excess) and that, if sustained, it could damage the germ plasm." Thus future generations could inherit the degenerative tendency. Yet consistent with the Protestant reform ethos of the successive periods known as the Gilded Age and the Progressive Era, the degeneration problem was not insoluble. "Whereas bad living could induce degeneracy," Rafter observes of this historical viewpoint, "clean living might reverse the process: in a few generations a family that carefully avoided vice might restore itself to normality."[35] Thus the cognitively dissonant interplay of biological determinism and unstinting confidence in social reform pervades the family studies literature treating criminality.

Edwin H. Sutherland and the Challenge to Hereditary Criminality

Not all criminologists were enamored of the eugenicists' analysis of criminality. One such critic, Edwin H. Sutherland (1883–1950), built his reputation as a criminologist during the 1920s. With the publication in 1924 of *Criminology* Sutherland was able to represent the environment-oriented reaction to hard hereditarian interpretations of criminality prevalent at the time. Sutherland was Assistant Professor of Sociology at the University of Illinois. In the 1930s he moved to Indiana University to found the Bloomington School of Criminology. In 1939 he served as president of the American Sociological Society. It was at this time he coined the term "white-collar crime." Rejecting biological determinism, and emphasizing the social connections and relationships of the criminal, Sutherland was

---

34. Ibid., 414.
35. Rafter, *Creating Born Criminals*, 36.

a key figure opposing hard hereditarian interpretations of criminality and eugenic social policies.[36]

The 1924 publication of *Criminality* laid out in seed form what would become Sutherland's fully developed Differential Association Theory of crime, filled out in the third edition of his *Principles of Criminology*. By 1924, Sutherland was convinced that the born-criminal theorizing of Lombroso and of many interpreters of Dugdale's Jukes family studies was fundamentally flawed. On the family studies, Sutherland pointed out methodological errors. "Every child in these families was subjected to the influence of environment as well as of heredity," he enjoined, "and the environment in almost every case was bad during the early formative period of life." He pointed out that "The appearance of a trait in successive generations does not prove that it is inherited." With sardonic satisfaction, he quipped, "There is the same logic in asserting that eating with a fork is inherited because it appears in successive generations as in asserting that criminality is inherited because it appears in successive generations."[37] He further criticized studies of Americans Heinrich Ernst Goring and Arthur Estabrook, and German Carl Rath, claiming that the offspring of criminals were themselves criminal according to a Mendelian ratio. Citing small sample size and circular reasoning in such research, Sutherland also noted that studies of criminal families lacked the methodological control routinely practiced in plant and animal studies. No strain of criminal family is sufficiently pure (i.e., free of environmental influences) to guarantee the hereditarian interpretation of multi-generational criminality.[38] Sutherland noted that most of the traits attributed by hereditarians to hereditary factors could be explained by environmental factors, social factors, abnormalities of intelligence, problems with the endocrine system, or temperament. Sutherland firmly asserted, "No one has such an heredity that he must inevitably be a criminal regardless of the situations in which he is placed or the influences that play upon him."[39]

Given this skepticism, it is little surprising that Sutherland opposed involuntary sterilization of the inappropriately identified "born-criminal." Having rejected the notion that criminality is inherited or heritable, Sutherland even questioned the use of sterilization for the "feeble-minded and insane." He questioned the claim that their mental defect was inherited. He stood unconvinced by the claim that mental defect predisposed such persons toward criminal activity. "Consequently there is no evidence that

---

36. Sutherland, "The Theory of Differential Association," 64–68.

37. Sutherland, *Criminology*, 113.

38. Ibid., 113–14.

39. Ibid., 117–18.

criminality would be reduced appreciably by a universal application of sterilization laws to the feeble-minded or insane," he wrote. Yet he did not completely shut the door on the practice. He owned that there could be other advantages to such laws beyond eugenic considerations. The reader wonders if Sutherland hedged his bet in light of the high degree of popularity of sterilization and the momentum of the eugenics movement in the early 1920s. He equivocated even further with the following words: "In so far as it can be determined that criminality is connected with an inherited abnormal trait, it is clearly desirable to stop the reproduction of people who have that trait." Citing Harry H. Laughlin of the American Eugenics Society, he added: "Probably all such persons should be sterilized and, in addition, some of them should be segregated if they are unable to care for themselves, or are for other reasons not safe in general society."[40]

Perhaps there is not the glaring inconsistency here that exists on the surface of these contiguous passages. Given Sutherland's previously (and oft-repeated) criticisms of hereditarian interpretations of complex social problems, his qualifying language, e.g., "in so far as criminality can be determined" as having a clear and direct connection to an "abnormal trait" gave Sutherland the room to question such determinism on a case-by-case basis. So he could remain open to the very popular regime of sterilization in principle, while trying to rein it in as practiced on any given individual.

It was not only criminologists who proposed solutions to crime. Sociologists also incorporated analyses of criminality into their generalist textbooks on sociology. An exemplar of a religious sociologist engaged in a struggle to come to grips with how to ameliorate crime in America was John Lewis Gillin.

John Lewis Gillin and the
Hereditary Criminal

John Lewis Gillin (1871–1958) was born in rural Iowa. He received the A.M. at Columbia University in 1903, and the B.D. from Union Theological Seminary in 1904. Gillin was ordained a minister of the Church of the Brethren. After acquiring a Ph.D. at Columbia in 1906, he served teaching stints at Ashland College (Ohio), and the University of Iowa. Appointed to the Sociology faculty at the University of Wisconsin, where he served until his death in 1958, Gillin worked alongside a giant in the field of sociology, Edward A. Ross.[41] Wisconsin was a hotbed of the ideology of government

40. Ibid., 621–22.
41. Rafter, *Creating Born Criminals*, 73, 89.

controlled social reform under the influence of academic experts, about which numerous books were published during the Progressive Era touting "The Wisconsin Idea."[42]

Albert T. Ronk, a denominational historian of the Brethren Church, whose sympathies gravitated toward the fundamentalist end of the fundamentalist-modernist continuum, contended that Gillin "was liberal in his theological thinking." Indeed, he cited Gillin to the effect that the Brethren Church ought to affirm that "Christianity is a life instead of a creed." Yet this historian qualified his assessment, noting that Gillin did not spend all his time "teaching liberal theology or social gospel." He further acknowledged that Gillin was "Brethren to the core and remained in the denomination" throughout his life.[43] Thus, though Gillin may properly be seen as standing on the leftward end of evangelicalism, his rhetoric evinced a lingering debt to the evangelical heritage of the nineteenth century.

In terms of Gillin's approach to eugenics, an analysis of Gillin's theological position must allow for his adoption of differing emphases in ecclesiastical and academic contexts. It must also allow for the likelihood that Gillin's theological leanings altered over time. His study of criminology and his study of theology indicated a continual dialectic, and mutual permeability. In Burger's estimation, "Gillin's criminology can be described as an attempt to unite the religious rhetoric of his predecessors with science." Gillin embraced the language of positivism, but retained his conviction that "only religion held the ideals that would enable the perfect state to be established." Science could go a long way toward ameliorating social ills by contributing knowledge of their causes and even solutions. But only religion, particularly Christ's ideals concerning the Kingdom of God and love of neighbor, could produce the impetus for the perfection of society.[44] Gillin resisted the deterministic theories of criminality of the highly influential Italian criminologist, Cesare Lombroso. He allowed for both hereditary and environmental influences in producing criminal behavior, and even emphasized the latter.

Gillin faulted the separation of capitalism from the Christian love ethic for much of the rise of criminality, yet called for reform, not rejection, of capitalism. In tune with the Social Gospel emphasis of many of his contemporaries, Gillin castigated ministers who supported the rich by failing to challenge exploitative practices toward the poor. Gillin spoke of salvation "as a matter of the here and now, as well as of eternity." He averred that "The religion of the Bible is a social religion." He clarified his interpretation of

42. Leonard, *Illiberal Reformers*, 40–42.

43. Ronk, *History of the Brethren Church*, 297.

44. Burger, *American Crime and Punishment*, 71.

social religion by defining it as "a religion having to do with man's relationships to others, both divine and human and supplying ideals and motives for the adjustment of his relationships to others in the interests of justice and right between himself and others." Gillin offered numerous concrete areas of social reform for greater church involvement, including milk purity, tuberculosis care, recreational facilities for the poor, assimilation of immigrants, and penal reform.[45] Burger's analysis concludes by affirming that "Gillin's criminology was one that actively sought a Christian state in America." Yet Gillin's positivism, not his religiosity, was the theme that came to predominate in the subsequent development of the field of criminology, and, according to Burger, "he was the last major writer to utilize God in his arguments." The arguments drawn from the secular realm and embraced by Gillin continued to influence later sociology, whereas his theological language did not.[46]

Gillin's rhetoric of Christian brotherhood, and his repeated emphasis on the love ethic of Christianity, rendered his opinions on eugenics all the more jarring. Gillin took positions on eugenics and utilized rhetoric every bit as harsh as hard-core eugenicists such as Charles Davenport. Though he did not embrace a hard-hereditarian etiology of criminality, he did embrace the negative eugenic solutions usually promoted most vigorously by hard hereditarian eugenicists.

A few citations from Gillin's major sociological and criminological texts bear out this contention. Addressing the question of the heritability of criminality, Gillin parted company to some degree with Cesare Lombroso and other hard hereditarian criminologists. "Crime as such cannot be inherited," Gillin asserted in 1926. Instead, he noted that "crime is a social phenomenon produced by a combination of the bodily and mental characteristics of the individual and the environment acting upon that responding personality." Recognizing the departure of the most recent genetic science

45. Gillin's role for the church imbibed the standard positivist line of dividing "facts" from "values." Gillin argued, in 1933, that "(T)he Church in its attempt to solve the problems of society, must take into account all knowledge that science may contribute. It must heed the economist, the sociologist, the psychologist, the biologist, the political scientist, and the educator. Its contribution must be social ideals, motives, and sanctions. To it is committed the responsibility of developing in the individual a social attitude and of seeking in every way to realize in social relationships the ideals of brotherhood, of kindness, of justice, and equally of opportunity. The scientist contributes his knowledge of the way in which things operate. The Church contributes its ideals of a righteous society and supplies the motives for the realization of these ideals." Gillin, *Social Pathology*, 534.

46. Burger, *American Crime and Punishment*, 71–89; see Gillin, "Study of Social Religion and Social Reconstruction," 103–105.

from the simplistic formulas of the more dogmatic eugenicists, Gillin noted that "Crime is not a unit biological characteristic." He did, however, acknowledge the role of heredity in the potential for crime, as "the natural characteristics which function in producing crime" which "may be inherited." In fact, Gillin afforded to heredity a significant role, if not as efficient cause, at least as a factor strongly correlated to crime:

> Therefore, while it is impossible to say that crime can be inherited, it is of the greatest value to study carefully the biological characteristics which are most closely connected with criminality in order to understand the genesis of the criminal and to know how to deal with the individual delinquent.[47]

Thus, what Gillin took away from hard hereditarian eugenicists with one hand, by a rhetorical sleight-of-hand he gave them back under the other by an even looser account of causality. Historians have rightly noted that the very sloppiness of such reasoning was precisely the cover under which numerous eugenic programs were carried out.[48]

Gillin cited with approval, and at length, Henry H. Goddard's studies of feeble-mindedness.[49] In 1926, Gillin was pragmatically cautious about the use of negative eugenical solutions such as involuntary sterilization. Noting that some states had passed sterilization laws, Gillin observed that "in the present state of the law and of public opinion, such laws applied to criminals have been declared unconstitutional as cruel and unusual punishments." Yet it was clear what solution Gillin would prefer, if the political climate would accommodate it:

> If, in the present state of public opinion, it is impossible to sterilize all who should not reproduce, and if on account of the cost it is impossible to segregate in institutions all the feeble-minded, nevertheless there are some things we can do to diminish criminality on the part of the feeble-minded.[50]

47. Gillin, *Criminology and Penology*, 159.

48. See Hasian, *Rhetoric of Eugenics in Anglo-American Thought*, 25–30, for a discussion of some eight distinct definitions of eugenics in the public discourse of this period.

49. Gillin, *Criminology and Penology*, 161–62; 170–76; cf. Goddard, *Kallikak Family*, 18–30; and Goddard, *Feeble-mindedness*, 145–46; 191–92; 233–34; 497–98; 516; 585–86. Goddard himself supported legalized involuntary sterilization, but much preferred segregation to sterilization. He also opposed sterilization as a punitive measure for rapists and other convicts. See Zenderland, *Measuring Minds*, 226–27.

50. Gillin, *Criminology and Penology*, 176–77.

Opting for the more optimistic approach to heredity and crime, Gillin endorsed "a program of positive eugenics," by which he apparently meant a program of educating the public about the child-like nature of the feeble-minded. "We shall not place upon them responsibilities in economic or social life which they are unable to carry," he explained, "We shall then throw about them the safeguards which we place about children." Acknowledging the role of environment, Gillin adopted a paternalistic stance, vouching that "The public attitude, instead of taunting them into criminality, will guard them and protect them from the temptations they are unable to withstand."[51]

In another book, also published in 1926, Gillin described a debate within the state of Wisconsin over the advisability of sterilization. In 1913 the Wisconsin Legislature passed a law empowering the State Board of Control to oversee screening of the feeble-minded for the purpose of sterilization. By 1918, some 150 inmates of the state's Home for the Feeble-minded had been sterilized.

The policy was not without its detractors. A Professor Guyer addressed the Wisconsin State Conference of Charities and Correction in 1913. He questioned acutely the ability of experts to determine which mental traits were hereditary. He poignantly asked, "Of the 27 or more recognized forms of insanity, who knows with any considerable degree of certainty which are heritable, which not?" Gillin laconically noted that Wisconsin's Board of Control resolved this problem "in a very sensible way." Gillin expressed satisfaction with the limitation of the law's application "to inmates of the Institution for the Feeble-minded," and "only to those whose family histories show an undeniable taint."[52]

Gillin, circa 1926, did at least allow that there were legitimate concerns about the policy of involuntary sterilization. It is clear, however, that he inclined toward a limited application of sterilization policy. In 1927, however, a significant national event occurred that doubtless influenced an important transition in Gillin's thought. In 1927 the United States Supreme Court issued its infamous *Buck v. Bell* decision permitting involuntary sterilization across the land.[53] Gillin's writings on sterilization thereafter show a marked hardening of attitude toward the problem of the feeble-minded.

51. Ibid., 177.

52. Gillin, *Poverty and Dependency*, 378–80.

53. 1927 marked a turning point in eugenical sterilization due to the *Buck v. Bell* decision whereby Virginia's sterilization law was upheld by the Supreme Court, 8–1. "This judgment greatly accelerated the passing of other sterilization laws. Sixteen states in the late 1920s and early 1930s passed such measures. *Buck v. Bell* also emboldened state officials to put existing laws into practice, causing a tenfold jump in the average annual number of compulsory sterilizations." Dowbiggin, *Keeing America Sane*, 78.

In *Social Problems* (1928), Gillin authored a chapter entitled "The Problems of Poverty." In this chapter, his earlier reticence or nuance on the question of involuntary sterilization had vanished. Now Gillin was enthusiastic about the practice, as evident from the following passage:

> The feeble-minded can be handled from the eugenic point of view only by some method that will prevent their reproduction. They can be segregated into institutions and colonies where the higher grades can make their living and some of them can be trained for certain vocations in which they can make their way on parole. Greater numbers could be let out into society after being properly trained if they were sterilized. A great deal of prejudice, however, exists against sterilization, and it is not making the headway it deserves. However, just recently the Supreme Court of the United States has decided that sterilization laws relating to the feeble-minded are constitutional. In this country we have attempted to take care of the insane by hospitals for those that may be curable and asylums for the chronic cases. No widespread effort has been made to sterilize the insane who have a history of hereditary insanity. This should be done or they should be kept in an institution where they cannot reproduce.[54]

For Gillin, the rights and prerogatives of the society took precedence over the rights of the individual, at least insofar as the object was the feeble-minded person. This was a common view in the heyday of eugenics. What seems remarkable is the tension between Gillin's draconian stance toward the feeble-minded on one hand, and his recurrent emphasis on love of neighbor as key to ushering in the Kingdom of God. Later in the same book, in a chapter entitled "The Church and Social Problems," this emphasis became explicit. "When we come to appreciate the importance of good health to the social welfare of our people and when we come to look upon the church as the agency by which the Kingdom of God can be brought on earth," Gillin intoned, "then we shall begin to test the vitality of the churches, not by the number in attendance, or their architectural grandeur, but, among other things, by the death-rate and sickness-rate."[55] In other words, pathologies such as persistent feeble-mindedness in the population could be interpreted as signs of the failure of the church to usher in the Kingdom of God. Elimination of such pathologies, even by the harsh methods of segregation and sterilization, were thus construed as nothing short of a divine mandate. By this logic, eugenical sterilization appeared to have the

54. Gillin, et al., *Social Problems*, 406.

55. Ibid., 484.

twin *imprimatur* of both religion and science. Others expanded the scope of sterilization to include not merely the feeble-minded criminal, but those of immigrant stock. One such figure was Chicago attorney David Orebaugh.

Keeping the Nordic Bloodstream Pure

David A. Orebaugh, a member of the Chicago Bar, authored the 1929 tome *Crime Degeneracy and Immigration: Their Interrelations and Interreactions.* His thesis was straightforward: ". . . the increase of crime in America has been due to a degeneration of the national stock resulting from its inter-mixture with disharmonic and inferior races," and "such degeneration will continue to spread at an accelerated rate unless it is grappled with in a de-termined and systematic way."[56] Class consciousness as well as bias toward Northern European (e.g., "Nordic") stock characterized his opening chap-ter: "It is significant and indicative of the moral and intellectual character of our early immigrants, and of the quality of the germ plasm transmitted to their posterity, that their numbers comprised but few of the degraded and subservient peasant classes."[57]

Orebaugh's nativist rhetoric was strongly tinged with disdain for the "melting pot" concept, as he decried *"dilution and pollution of the hereditary stream by inferior alien heritages."*[58] Citing the influx of eastern and south-ern European immigrants after the Civil War, mainly as cheap labor, Ore-baugh lamented that "this blood now constitutes 70 percent or 75 percent of the total immigration." He added that "we find that we are getting progres-sively lower and lower types from each nativity group or race."[59] Orebaugh warned that from these populations "with their meager mental equipment, the product of centuries of degraded inheritance . . . our criminal element is largely recruited."[60]

It would be rather simple to dismiss Orebaugh as just another xeno-phobic crank. The problem with this analysis is that Orebaugh was able to cite approvingly similar warnings from professors of major American uni-versities, including Ivy League schools. What is remarkable is the degree of visceral revulsion evident in Orebaugh's rhetoric, some five years after the enactment of the Johnson Immigration Act of 1924, the most restric-tive immigration policy initiative in U.S. history. The act had rolled back

56. Orebaugh, *Crime Degeneracy and Immigration*, vi.

57. Ibid., 4.

58. Ibid., 5, emphasis in original.

59. Ibid., 10–11.

60. Ibid., 14–15.

immigration quotas to 2 percent of 1890 levels. This was meant to dramatically curtail such numbers for eastern and southern European immigrants. Clearly Orebaugh feared the interbreeding of those of Nordic stock with others of allegedly inferior stock among the recent immigrants already in the U.S. His etiology of criminality was stark: "Racial dilution produces degeneration; degeneration predisposes to crime; criminals practically without exception exhibit some of the stigmata of degeneration; hence, racial degeneration may be said to be a cause of crime and criminals."[61] Historian of immigration, Mae M. Ngai notes the following about the Johnson Act: "At its core, the law served contemporary prejudices among white Protestant Americans from northern European backgrounds and their desire to maintain social and political dominance. Those prejudices had informed the restrictionist movement since the late nineteenth century."

Yet prejudicial thinking was certainly not a new phenomenon. The question is, what new strain of prejudice was showing evidence of a special virulence. Ngai continues: "But the nativism that impelled the passage of the act of 1924 articulated a new kind of thinking, in which the cultural nationalism of the late nineteenth century had transformed into a nationalism based on race."[62] Biology, filtered through the distorting lens of eugenics, was shaping public policy in a dramatic way.

Orebaugh devoted an entire chapter to degeneration theory, citing B.A. Morel's emphasis on degeneration as "*morbid deviation from an original type.*" Orebaugh self-consciously expanded Morel's definition "to include all morbid deviations from conventional standards of life, conduct and mentality as are peculiar or habitual to the individual, and clearly obvious to the average normal intelligence."[63] Thus, any unconventional personality could be deemed degenerative under Orebaugh's expansive new definition. Whereas Morel the psychiatrist was engaging in research on mental patients, Orebaugh the attorney interjected general morality, class, and nationality into the degeneration equation. Thus degeneration language became a wax nose, a catch-all moniker for all who departed from the undefined norm.

Basing his remarks on the insights into heredity from horticultural research, Orebaugh averred that nature would eliminate the inferior when not interfered with. But human intervention could tip the balance in favor of the inferior, such that "progressive evolution is arrested and retrogressive evolution is substituted." Approvingly citing raceologist extraordinaire Madison Grant, Orebaugh decried cross-breeding of superior and inferior

---

61. Ibid., 20.

62. Ngai, *Impossible Subjects*, 23.

63. Orebaugh, 34–35; emphasis in original.

stock. Orebaugh warned that: "A fact apparently well attested in science is that superior traits and qualities are recessive and tend to disappear from the stock when imposed upon an inferior strain by crossing, while at the same time the retrogressive qualities attain to dominance."[64]

In what sense then, given this topsy-turvy logic, could recessive traits be dubbed superior? Mendelian notions of dominant and recessive traits certainly did not demand or justify such a prejudicial application to human social organization. Given the authority of science, however, Orebaugh was compelled to invoke it and apply it to his social policy agenda. Marriage restriction was in view when he cautioned that "the finest human specimens" if mated with "inferior, anharmonic or defective persons," would yield at least some "progeny having the weaknesses and defects of the inferior parent." He warned that "even the superior offspring of such union will carry in their germ plasm a sufficient number of *genes* or *determiners* to insure the transmission to some of their own children of the inferior qualities of their defective parent or grandparent."[65] He concluded that: "Heredity, education and tradition all are factors" in social development, "but to heredity must be ascribed the greater influence." Orebaugh held that it was the "quality of the blood-stream" rather than cultural factors transmitted from one generation to the next, "that determines the real moral and spiritual worth of a people."[66]

Orebaugh was concerned less with physical degeneration than "the mental, moral, and spiritual decadence produced by degeneration." Such would explain both minor deviations from social norms as well as "the degeneracy which pre-disposes to crime in its more obvious aspects and manifestations."[67] Degeneration theory had become a totalizing discourse, an explanation for every degree and variant of perceived deviancy. It brought together distortions of both science and religion, and became a tool in the reaffirmation of lingering prejudices and fears that seem always to look for a scapegoat. James A. Morone argues that this has been a pervasive intergenerational theme in the American ethos. He writes: "If moral fervor stirs our better angels, moral fever spurs our demons. Frightening changes . . . rouse fears of decline. Every generation blames a slack-virtued, un-American 'them.'"[68] True to this ethos, degeneration theory, with both

---

64. Ibid., 36.
65. Ibid.
66. Ibid., 225–26.
67. Ibid., 37.
68. Morone, *Hellfire Nation*, 3.

its scientific and religious resonances, was for a time a powerful tool in the toolbox of social control.

The protean nature of degeneration can be seen in Orebaugh's identification of alleged by-products of degeneration. He saw in degeneration the fertile soil for such disparate social phenomena as: yellow journalism, disintegration of parental authority, the waning of school discipline, the exaggeration of sex, fetish worship, inebriety and drug addiction, the coddling of criminals, and a growing disrespect for law. In progressive fashion, however, Orebaugh believed that America could avoid the degenerative pattern of the fall of past civilizations. By its accumulated knowledge, the body politic could diagnose and counteract its degenerative weaknesses.[69]

Orebaugh's proposed solution to the problem of degeneration was no less sweeping than the diagnosis had been. He called for expansive social reforms to alter dramatically the face of the nation. Going even beyond what most eugenicists lobbied for, Orebaugh's vision teetered on the edge of utopianism. First, and unsurprisingly, he called for further immigration restrictions. His vehemence for such restrictions can be measured against his fear that the forces of repeal of the Johnson Immigration Act might succeed. Citing a set of hearings in 1927 designed to roll back the restrictions, Orebaugh feared the dam against a flood of non-Nordic stock might be cracking. He called for "the assimilation of the large number of aliens already here and the maturing of a homogenous, like-minded population before we let down the bars to fresh hordes."[70] Failure to uphold restrictions would lead to racial degeneration, disintegration of American ideals to the level of Latin America, leading to an increase in crime. Orebaugh called for "radical steps to prevent that association and propinquity of races which always results in intermixture and degeneration." He pled: "They must keep the Nordic bloodstream pure and undefiled by excluding all inferior and disharmonic elements."[71]

Further measures included standard themes debated in the eugenics literature, such as "the sterilization of defectives," birth control, and the repeal of old statutes forbidding artificial birth control methods. Drawing on his legal background, Orebaugh went on to argue for pardon, probation, and parole, and for the replacing of many jury trials with trials by judges. This was in part due to his belief in the lack of intelligence of the average juror due to the ravages of degeneration. Along the same elitist line, Orebaugh favored the appointive judiciary over the elected judiciary. In fact,

69. Orebaugh, *Crime Degeneracy and Immigration,* 43–89.

70. Ibid., 115, 118–21

71. Ibid., 132–33.

Orebaugh called for many democratically elective positions to be replaced with appointed positions, again out of fear and distrust of the mental inferiority of the masses. He believed replacing aldermen with commissioners would rein in corruption. In essence, a meritocratic bureaucracy would replace the democratically elective system, with its corruptive deficits.[72]

Orebaugh called for educational reform, in which "the evils resulting from the interbreeding of disharmonic races should be explained and illustrated by examples from the vegetable and animal worlds." Further, "eugenics and genetics should be made studies of first importance."[73] He called for reforms in the press, the home, the church, and other agencies. Orebaugh appealed to Reformation and Protestant churches, not the church "with its traditions of medievalism and reactionism" in an obvious swipe at Roman Catholicism. He chided this institution for closing its mind "against the teachings of science," preferring traditions over "the obvious laws of the universe" and "the enlightened reason of mankind." Instead, he urged the church to change such attitudes in order to "take and keep its rightful place as an agency in the regeneration of the race." Further, he urged the church to "strip away the unessentials of dogma and creed, of supernaturalism and ritual," instead to work harmoniously with science. In this way the church would be able to affirm "the fatherhood of God and to realize the brotherhood of man" while "at the same time actively sponsoring, aiding and promoting the specific program for race betterment," as would befit its "true mission on earth."[74]

Conclusion

As the Great Depression descended on the U.S., many who had been the elites found themselves in poverty. Environmental factors in the loss of former status and health became more obvious to many who had lived a life of privilege, a life to which they felt a sense of hereditary entitlement. Further, the biggest criminals, it turned out, had been the elites who held power over the banking and finance industries and oversaw their collapse to the harm of many less fortunate than they. In the thirties, anthropologists began to expose just how powerful culture was, and began to rein in the excesses of hereditarian explanations of social problems. Further, the willingness of the much-vaunted German intellectuals to take eugenics ideology to horrifying conclusions began by the late 1930s to chasten many

72. Ibid., 134–211.

73. Ibid., 232.

74. Ibid., 234–36.

American eugenics enthusiasts, or at least to curtail their public rhetorical excesses. Others have noted the obvious athletic prowess of Jesse Owens at the 1936 Olympic Games in Berlin as a clear implicit rebuke to eugenic doctrines of Nordic supremacy.

Both religious and non-religious elites in the Gilded Age and in the Progressive Era agreed that the rising crime rate bespoke an intractable problem in American society. Hereditary criminality relieved some of the pressure on social reformers for the seeming failure of environmental reforms to reduce crime. After all, if a certain percentage of criminals were criminals because of the relentless workings of heredity, then their recidivism and resistance to reform were not the fault of their minders, but of impersonal forces many generations in the making. Indeed, in a strange way this should have led to a deeper compassion toward the criminal. Perhaps he or she was just a hapless victim of forces outside his or her control. A few professionals played this theme up with greater consistency, such as famous attorney Clarence Darrow, who defended accused persons overtly using just such deterministic logic. The moralizing rhetoric of most social reformers, however, combined with their desire to label problematic citizens, including those accused of crimes, in a scientific way, seemed anything but humane.

The criminalization of the sale of alcohol was another way that social reformers sought to slow the degeneration of American society and Western civilization. Indeed, the theme of alcohol abuse or "inebriety," with its doleful societal effects, dominates historiography of the early decades of the new century. The history of social control and social reform in this period has few narratives more powerful than that of the move toward the prohibition of alcoholic beverages. How that story intersected with the science of heredity and with the Protestant ethos of temperance forms the subject of the next chapter.

CHAPTER 8

Drink and the Degeneration
of the Germ Plasm

[H]ereditary effects in man, and, therefore, racial deteriora-
tion, can follow either from single intoxications or from chronic
alcoholism.

—Cora Stoddard, 1925[1]

Introduction

DEGENERATION THEORY WAS A major influence on late-nineteenth and
early-twentieth-century theories on the causes and potential cures for alcohol abuse. In her major study of the social construction of alcoholism
as a disease in this period, Sarah W. Tracy notes that "Aberrant behavior
such as chronic drunkenness was often explained through recourse to
vague somatic causes such as neurasthenia and hereditary degeneration,
but overindulgence, overwork, poverty, and disappointment remained oft
cited causes of alcoholism."[2] Tracy describes Judge L. G. Kinne, a powerful
member of the Iowa Board of Control of State Institutions as an example of
this mentality. In 1904 Kinne relied upon degeneration theory to argue state
institutions ought to offer treatment to the inebriate members of the populace. The Board of Control dreaded the results of failing to house and treat
alcoholics with degenerative tendencies. The financial factor loomed large
as a motivation, and Tracy avers that the state's public servants feared that
". . . if no efforts were made to confine and treat the inebriate ranks of Iowa,
this dangerous subpopulation had the potential to spawn a race of physical

1. Stoddard, "Alcohol," 1:118–24.
2. Tracy, *Alcoholism in America*, xii.

and moral degenerates that would tax the state and national coffers as never before." The interests of the state converged with scientific interests, in that "inebriate reform was promoted as enlightened and scientific statecraft."[3]

As Daniel Okrent has noted, in the buildup to Prohibition many claims were made for the temperance cause in the name of science that, both then and now, stretch credibility. Such claims were often made by the Woman's Christian Temperance Union (WCTU) under the rubric of "Scientific Temperance Instruction." Some examples include claims that more than half of those who imbibe beer will die from dropsy, or that alcohol burns off the skin of the throat when swallowed. Similar declarations showed up in approximately half the public school textbooks in America by 1901, and temperance instruction had become mandatory three times per week in every state.[4]

The 1917 edition of the influential *Cyclopedia of Temperance Prohibition and Public Morals* contained an article on heredity. After citing statistic after harrowing statistic from European studies on the birth defects and developmental abnormalities ostensibly caused by the drinking of alcohol, the study moved to the subheading "Some American Findings." Noting several studies of alcohol's effects on dogs and guinea pigs over multiple generations, the authors moved to cadaveric research on human alcoholics, warning: "Very recently autopsies upon drinkers have revealed that when tuberculosis and similar diseases have failed to cause atropy of the testicle and otherwise injure the reproductive power, the constant consumption of alcohol has the power to do so."[5] The implication was clear: such an evocative visual image was intended to awaken fear over the damage being done to future children and future generations symbolized by that unfortunate and now ineffectual organ. The article went on to cite a New Jersey Commission finding that even moderate drinking "Is the cause of a great majority of the epileptic, feeble-minded, and subnormal children in that State."

Some of the science of the era has endured, such as the recognition of the devastating effects of fetal alcohol syndrome on the gestating child in the womb. Other aspects of the malleability of the era's hereditary science represented degeneration theory as the adjunct of some strange prenatal speculations that blended temperance fervor with folk science. For this we must turn to the nineteenth-century connections between degeneration theory, with its blend of moralism and hereditary research, and alcoholism as a social problem.

3. Ibid., 210.

4. Okrent, *Last Call*, 21.

5. *Cyclopedia of Temperence Prohibition and Public Morals*, "Heredity," 187–89.

Temperance and Nineteenth-Century
Views of Heredity

For social reformers of the late nineteenth century, characteristics could be acquired (echoing Lamarck's theory of inheritance of acquired characteristics) yet also could become heritable in an intransigent manner (thus reflecting the hard hereditarianism of Weismann). How these characteristics affected future generations in a downward spiral of debilitating heredity was also articulated in terms of Bénédict Morel's degeneration theory, an inheritance of its wide acceptance in European medicine and psychiatry, and as imported into those fields the United States.[6] In 1889, Michigan Jurist Sanford Green repeated this anecdotal warning from Morel's research with reference to alcohol abuse:

> Dr. Morel gives the history of another family, in which the great-grandfather was a drunkard and died from the effects of intoxication, and the grandfather, subject to the same passion, died a maniac. He had a son far more sober than himself, but subject to hypochondria and homicidal tendencies, and the son of the latter was a stupid idiot. Here we see in the first generation alcoholic excess; in the second hereditary dipsomania; in the third hypochondria; in the fourth idiocy and probable extinction of the race.[7]

Such anecdotal evidence was remarkably commonplace in the decades following the dissemination of Morel's degeneration theory. One highly popular method of extending the insights of degeneration theory was the family studies literature, surveyed more fully in chapter 4. The multi-generational impact of alcoholism was an existential reality in the families who suffered its effects. But in a scientific age when biology burgeoned in its ability to explain complex social phenomena in simple terms, alcohol abuse became increasingly the province of physicians and psychiatrists, and then of legislators, judges, and penologists. Clergymen also served on boards related to these fields, and many agitated for prohibition of the sale of intoxicating beverages, first state-by-state, and then nationally, culminating in the 18th Amendment, commonly known as Prohibition, ratified January 16, 1919. The litany of unintended consequences of this amendment and its enforcement legislation is well-documented. Still, the harmful effects of alcohol abuse fell disproportionately on women and children, and such cannot be responsibly ignored as part of the story of temperance agitation. As Okrent

6. An excellent overview is Bynum, "Alcoholism and Degeneration," 59–70.

7. Green, *Crime*, 28.

points out, in 1840 Americans consumed 36 million gallons of beer; by 1890 that number was a staggering 855 million gallons. Indeed, the population tripled over that period, but beer consumption (not even factoring in a thriving trade in hard liquor) had expanded twenty-four-fold.[8] Those who bore the heaviest burden of this increase tended to be the women and children of the USA. The main organization that arose to give voice to their plight was the Woman's Christian Temperance Union.

WCTU and the Crusade for Purity

The Woman's Christian Temperance Union was a force to be reckoned with in more social policy arenas than prohibition. Most histories of the WCTU, not surprisingly, have focused on its role in the movement for prohibition, sometimes to the neglect of its other cultural reformist agendas. One such agenda was social purity or hygiene, employing a range of ideas indebted both to nineteenth-century notions of heredity and to the emerging eugenics movement. While at first glance it might seem that eugenics would have been anathema to evangelicals such as predominated the WCTU, upon close examination the opposite proves to be the case.

Noting the plasticity of eugenic thought, Leila Zenderland has pointed out that eugenic ideals were popular among American charity societies, anti-vice initiatives, and women's clubs. In the late nineteenth century the WCTU sponsored "heredity meetings" at which WCTU member Dr. Jennie M'Cowen warned, in biblical language, against the heritable effects of sexual misconduct.[9] Temperance and social reform advocates selectively adopted degeneration language from both religion and science in efforts to curtail behavior detrimental to their civilizing mission.

Founded in 1874, the WCTU became the largest women's organization of the Victorian era, numbering 200,000 members by 1892, and over 344,000 by the year 1921.[10] The watchword for the WCTU became "home protection," and members worked for women's suffrage under that rubric. While these women often supported conservative and traditionalist assumptions concerning the roles of wife and mother, they also participated in activities that helped blur the lines between private and public spheres. Frances Willard, founder of the WCTU, oversaw an eventual nineteen departments, "each one devoted to achieving a specific social-reform goal,

---

8. Okrent, *Last Call*, 25–26.

9. Zenderland, "Biblical Biology," 512–15.

10. Parker, *Purifying America*, 5.

ranging from child-labor laws to international peace, and from arbitration to social purity."[11]

In 1881, the WCTU founded the Department of Scientific Temperance Instruction. This powerful political arm of the WCTU was capably headed by Mary H. Hunt from 1880 to 1906. Hunt and her highly motivated volunteers worked to persuade state legislatures across America to make scientific temperance education mandatory, first in the secondary schools, and later in the elementary schools.[12] Hunt's own account of the specific mission to get scientifically credentialed anti-alcohol textbooks into wide publication, distribution, and mandatory use in the public schools bears testimony to the authority of science in the 1880s. Her account of the first ten years of lobbying for scientific temperance instruction included the following multi-pronged rationale:

1. Public schools "must teach with no uncertain sound the proven findings of science, including: (a) "That alcohol is a dangerous and seductive poison;" (b) That drinks containing alcohol are thus "dangerous drinks, to be avoided, and that they are the product of a fermentation that changes a food to a poison"; and (c) Even a small amount of alcohol creates an uncontrollable appetite for more.

2. Schools must instruct students "the effect of these upon 'the human system,' that is, upon the whole being-mental, moral, and physical." This instruction should warn students about "the appalling effects of drinking habits upon the citizenship of the nation, the degradation and crime resulting," and must stress "the solemn warnings of science on this subject."

3. The instruction should be grade appropriate and understandable to the age of the child.

4. Temperance was to be the paramount topic, and scientific hygiene instruction should only be enough to promote temperance, yet a minimum of twenty pages should be given to "the question of the danger of alcoholic drinks and other narcotics."

5. Such textbooks should avoid the topic of the medical uses of alcohol.[13]

11. Ibid., 5–6.

12. Kerr, *Organized for Prohibition*, 49.

13. Hunt, *History of the First Decade of the Department of Scientific Temperance*, 35–36. A probative secondary text on Hunt's efforts is Zimmerman, *Distilling Democracy*, 15–38.

While such dry efforts were strongly religiously motivated, Hunt and her department repeatedly sought out credentialed scientific authorities to lend added legitimacy to the educational work of the WCTU. Such efforts bore tacit testimony to the cultural authority of the scientific enterprise at the end of the nineteenth century. Religious arguments, strategies, and institutions were necessary but not sufficient to advance and broaden the cause of temperance and eventually prohibition in the USA.

Given the negative influence of alcohol on both the home life and public morals of the nation, the WCTU promoted sexual purity as part of its overall social mandate. Alison M. Parker has delineated the various ways in which women of the WCTU promoted public purity through various campaigns of censorship. She has sought to show that such efforts were not merely limited to the WCTU activists, but were consonant with a widely shared, middle-class ethos in North America at the time. In an effort to protect the home from "impure" images and thoughts, the WCTU established departments devoted both to suppression of indecent literature, plays, and movies; and to promotion of alternative "pure" media. It increasingly lobbied for government involvement in such efforts, leading to various and oft-challenged censorship laws.

On the one hand, the WCTU promoted an increasingly centralized and interventionist view of the federal government for the protection of women and children. On the other hand, the WCTU maintained many of the conventional views of motherhood and home life held by nineteenth-century revivalist evangelicalism. Religious historian Mark Noll has commented that: "Overwhelmingly, Protestant women reformers favored an approach that was both loyal to a traditionalist Protestant Scripture and eager to reinterpret it for the sake of equity for women."[14] This bridging of the liberal/conservative divide by women's groups like the WCTU should qualify any model of Progressive Era Protestantism that tends toward a simplistic bipolarity of conservative versus liberal proponents.[15]

One way the WCTU rendered its message in progressive terms was to invoke scientific studies whenever such could be construed as bolstering the WCTU purity crusade. In 1924, the golden anniversary of the WCTU, Elizabeth Putnam Gordon described the expansion of Frances Willard's vision for the organization. The WCTU had subdivided its reform agenda into five general arenas: "Preventive, Educational, Evangelistic, Social and Legislative." In referring to the group's focus on far more than the single issue

---

14. Noll, "The Bible, Minority Faiths, and the American Protestant Mainstream, 1860–1925," 214.

15. See Harrell, "Bipolar Protestantism," 23–24.

of alcohol abuse, Gordon averred that "A scientific age required study of this subject in its correlations; and Frances Willard's plan allied the WCTU with all other moral forces." The WCTU initiative for catechizing America's children on the evils of alcohol was carried out under the rubric "Scientific Temperance Instruction."[16]

The WCTU was well poised to take advantages of newly emerging social sciences, with their wide-ranging studies of social ills and their correlates. Parker has pointed out that:

> The WCTU translated its relatively conventional maternalist concern for youths into sympathy with the Progressive child-study and hygiene movements and with the field of psychology. Evangelical WCTU members were well within the progressive cohort and did not share the hostility to the intellectual and scientific realm of the post-World War I fundamentalists.[17]

The women of the WCTU took a selective approach to the science of their era. Blanche Eames was merely voicing a common Protestant truism when she wrote "Science and the Scriptures, rightly interpreted, never conflict; they can not, (sic) because they are both the work of God; one His expression in Nature, the other in His written word."[18]

WCTU literature evinces a strong linkage between WCTU spokeswomen and the pronouncements of scientific eugenicists promoting mental, moral, social, and racial hygiene or purity. For the WCTU, this purity was not merely a matter of individual sanctity, however, but a requisite social precondition for ushering in Christ's kingdom on earth. For the WCTU, progressivism meant progress toward a divine ideal, not a mere earthly utopia. In promoting such a high-stakes vision of cultural transformation, WCTU leaders drew upon any and all persuasive means. WCTU representatives employed the language of professionals such as natural scientists, doctors, social scientists, and psychologists. WCTU members cited eugenicists such as Havelock Ellis and David Starr Jordan, who were leading voices of the social hygiene and eugenic movements. Social hygiene "provided moral reformers with a quasi-scientific language and professionally accepted metaphors of disease to describe the dangers of 'immorality.'"[19] Jordan was a favorite source for WCTU arguments, as he linked the liquor

16. See Gordon, *Women Torchbearers*, 35; and Zimmerman, "When Doctors Disagree," 171–97.

17. Parker, *Purifying America*, 9.

18. Eames, *Principles of Eugenics*, 14.

19. Parker, *Purifying America*, 23.

trade with such socially and morally unsanitary phenomena as prostitution and contagious diseases.

WCTU purity literature shows the profound dependence of WCTU arguments upon the science of their day—particularly the early stages of the science of heredity. Yet their arguments illustrate the confused state of the nascent science of heredity in the early twentieth century. Confusion about what constituted a scientific claim during this period thus should not be wholly unexpected. Historian of science Peter Bowler has shown that this era was especially tumultuous and unsettled when it came to the science of heredity. The period saw the birth of genetics as a discrete field within biological science, a birth at times painful and by no means quick.[20] Thomas C. Leonard has described this period as "plural and unsettled on fundamental questions: whether environment affected heredity, whether natural selection impelled evolutionary change, whether the individual or the group was the principle unit of selection, whether fitness consisted solely of reproductive success, and whether evolution offered progress or merely change."[21] Social reformers promoted social progress, and did not merely accept nature as purposeless change.

The inheritance of acquired characteristics view of heredity exhibited a protracted hold on the minds of non-specialists or laypersons including the reformers of the WCTU. The confusion in the public mind about heredity was only compounded when specialists themselves carried out sometimes-bitter debates over the precise effects of biological inheritance. Bowler has noted that belief in the inheritance of acquired characteristics was a tenacious and persistent idea:

> The inheritance of acquired characteristics is, of course, repudiated by modern genetics, because we see no mechanism by which the genes can absorb information from the body in which they are enclosed. Yet it has been accepted as a genuine effect since time immemorial and even today many non-scientists almost instinctively believe that the effect ought to work. The inheritance of acquired characters originated with a pre-Mendelian notion of heredity that is more in tune with common-sense or folk belief.[22]

Only by appreciating the real state of flux and confusion within the scientific community itself during this period can we appreciate how an even greater confusion and ambivalence over eugenic ideals obtained among

20. Bowler, *Mendelian Revolution*, 21–45.

21. Leonard, *Illiberal Reformers*, 97.

22. Bowler, *Mendelian Revolution*, 38.

reform-minded evangelical clergy and other non-specialists. The balance of this chapter is given over to contextualizing seemingly bizarre claims within a situation of fragmentation in both scientific and Protestant religious communities.

The Role of Frances Willard
in the Purity Crusade

In 1890, the indefatigable champion of temperance, women's suffrage, and a host of other social reforms, the Methodist Frances Willard (1839–1898) published one of the key speeches she had given repeatedly in the 1880s. Attacking the double-standard in matters of sexual purity, Willard defended the Woman's Christian Temperance Movement's "organized and systematic work" in the group's efforts "for the promotion of Social Purity."[23] Reaching into many homes to ameliorate a host of ills bred by alcohol abuse and male promiscuity, Willard boasted "Mothers' Meetings are becoming one of the most familiar features of the WCTU." Emphasizing the importance of pure over impure reading matter, Willard reported that "for these we prepare programs, leaflets, and courses of reading . . . from which hundreds of thousands of pledges and pages of literature have gone, as pure and elevated in style and spirit as consecrated pens could render them."

Willard commended the matrons of the WCTU who visited police stations for the purpose of reaching out to "arrested women." She also noted with approval that "state care for moral as well as mental incapables is being urged and with some small beginnings of success."[24] By promoting chastity prior to marriage for males as well as females, and fidelity therein, by co-education, and by elevating women's ability to exercise greater discernment and freedom in the choice of a mate, Willard believed a dismantling of the double standard would benefit the entire family unit. Willard also defended the right for a woman to remain "a maiden *in perpetuo*," and offered as a rationale for this principles of autonomy, and the good of posterity. Willard urged that allowing a woman greater freedom in marital choices forestalled

23. Willard, "A White Life for Two," 163. Giele has noted four major contributions of the WCTU to later feminism. First was an ideology connecting the influence of women in the domestic sphere to their growing strength in the public sphere. Second, the WCTU raised up prominent and highly visible women leaders. Third, the WCTU was a remarkable social movement organization, spanning the nation via its local chapters and providing a communications network for reforming women. Fourth, the WCTU reform strategy merged temperance with women's rights, thus creating a popular movement. See Giele, *Two Paths to Women's Equality*, 63–64.

24. Willard, "A White Life," 165.

the calamity of her marrying "a man whose deterioration through the alcohol and nicotine habits" was "a deadly menace to herself and the descendants that such a marriage has invoked."[25] Following a thinly veiled plea for
birth control freedom, Willard appealed to science and religion with equal
fervor, writing that "the study of heredity and pre-natal influences is flood-
ing with light the Via Dolorosa of the past; the White Cross army with its
equal standard of purity for men and women is moving to its rightful place
of leadership among the hosts of God's elect."[26]

In her use of biblical rhetoric, including tropes of her Wesleyan religious tradition, Frances Willard was the harbinger of a social reform crusade
increasingly reliant on the science of human heredity. In other figures in the
movement for whom the purity issue was paramount a mosaic of hereditary
notions came into play with both scientific and religious rhetoric blended.

Purity Evangelism and the Rhetoric of Eugenics

Mary E. Teats served as National Purity Evangelist for the WCTU at the
turn of the century. She was a lecturer for the National Purity Association,
and a lecturer of the Correspondence School of Gospel and Scientific Eugenics.[27] She attacked impurities of varying kinds often evincing a mindset
derived from a blend of biblical proof-texting, degeneration theory, and
Lamarckism.

Based in Oakland, California for much of her career, Teats traveled
widely throughout the U.S., giving hundreds of lectures and distributing
thousands of pages of literature for the WCTU. The 1895 *Minutes of the
WCTU* described her work for the previous year as follows:

> Eight sermons, besides others not reported. Five thousand
> leaflets. Good work accomplished in all lines. Railroad superin-
> tendents interested for pure literature. Seven thousand pages of
> Temperance and Purity Leaflets distributed. Young men influ-
> enced to read purer books.[28]

---

25. Ibid., 169.

26. Ibid., 170. The "White Cross Army" refers to a lapel pin to be worn by men
who made a public pledge to break the double standard by maintaining the same sexual
purity standard that Victorian society demanded of women.

27. Teats, *Way of God in Marriage*; and Eames, *Principles of Eugenics*, 87.

28. *Report of the National Woman's Christian Temperance Union Twenty-Second Annual Meeting*, 303, Willard Memorial Library Archives.

Her 1906 marriage manual, *The Way of God in Marriage* carried the cause of marital purity even further than her tracts and speeches could. This tome exemplified her effort to weave scientific and biblical authority together into a seamless argument. Teats averred in the opening chapter that "My sole object is to give what I firmly and conscientiously believe to be Divine truths, when God's word is allowed its highest interpretation" on the subject of "God's Way in Marriage." Because of the theological and social conservatism of much of her readership, she had to develop a rhetorical style that could be both self-effacing and still authoritative.

The WCTU generally worked within the traditional marriage paradigm of Protestantism, yet sought to shape marriage in unconventional ways. To accomplish this, Teats had to construct an authoritative voice for herself from within a shared authority-structure. She did so by appeals to scripture and to science. The self-effacing element of this paradoxical hermeneutic is illustrated by the following passage:

> While I lay no claim to being a "prophet," yet, with the light God has given me, I have endeavored to present in this work the unpolluted Word of God, from Gospel and scientific standpoints. Neither do I lay any special claim to being versed in the sciences. However, I have studied the science of life, the laws of the conservation of vital force, and the procreation of the human race, sufficiently to believe that the greatest tragedy of humanity has been the almost universal violation of God's laws in regard to the bringing into being of a human soul.[29]

Her appeal was to two sources of authority foundational to Protestant evangelical thought: scripture and science. Ministers formed arguably the foremost readership targeted by Teats, as evidenced by her continual use of anecdotes involving ministers, and her use of direct address to ministers.[30] In the absence of a clerical ordination or official ecclesial position, Teats appealed to the Bible, the final authority for virtually every Protestant denomination. Because Protestant theology allowed to all believers without distinction the right of interpretation, ministers could hardly object to such a procedure, even if they disagreed with her interpretations. To forestall such disagreement she went the further step of appealing to modern science whenever possible to bolster a particular interpretation or application of the sacred text to social ills.

Early on, *The Way of God in Marriage* laid out a line of interpretation the author would follow throughout its pages. In short, her argument

29. Teats, *Way of God in Marriage*, 2.

30. Ibid., 61–72; 84–85; 100–101; 138.

was that marital sexual incontinence has been, and will continue to be, the source of a host of societal ills unless checked by conscientious married believers. For Teats, sexual intercourse within marriage was to be reserved only to those times when both parents were in agreement that the goal of intercourse was to produce a child. This belief showed marked affinities with the tenets of the eccentric nineteenth-century marriage reformer, John Humphrey Noyes. Noyes had edited a journal aptly titled *The Perfectionist*, and had developed a community for practical application of his utopian experimental ideas on human sexuality. In his estimation, the *amative*, or social function of human sexual intercourse took precedence over the procreative function. Noyes urged sexual congress without "crisis" (i.e., ejaculation), along with communal living and "complex marriage." Noyes perceived his version of human sexual praxis as the best of the available means of avoiding unwanted procreation, so was an early attempt at birth control. Yet an important component of this view was his conviction that "the useless expenditure of seed certainly is not natural."

To Noyes, non-procreative emissions represented a violation of the divine purpose of sex. He wrote, "God cannot have designed that men should sow seed by the way-side, where they do not expect it to grow, or in the same field where seed has already been sown and is growing; and yet such is the practice of men in ordinary sexual intercourse." In other words, sexual intercourse, either without the intention of conception, or transacted during an existing pregnancy, violated both nature and the divine design.[31] The women of the WCTU, such as Mary E. Teats, held far more conventional views of marriage than did Noyes. But the same perfectionist impulse, coupled with a frank distaste for human sexuality, drew careful lines of distinction between the spiritual and sensual dimensions of sexual activity. This approach can be discerned in the efforts at marriage reform of both Noyes and Teats.

Teats regarded with particular abhorrence those couples who had intercourse during pregnancy. She believed the waxy substance with which many children come out of the womb was actually the residuum of spermatazoa illicitly expended in such conjugal unions. The source of this belief is not wholly certain, but its similarity to the science of stirpiculture, as the sexual ideology of Noyes came to be called, at least suggests a connection. To this should be added a combination of folk science, tendentious readings of biblical texts, and the influence of literature railing against the Victorian double standard in matters sexual.

---

31. Cited in McClymond, "John Humphrey Noyes," 225.

In *The Way of God in Marriage*, beginning with Genesis, Teats blended a qualified acceptance of the theory of evolution with an oblique reading of scriptural texts to buttress her claims. While briefly acknowledging that "Creation and evolution are both alike of God," Teats cited familiar phrases from Genesis, such as "Be fruitful and multiply and replenish the earth, *and subdue it*" (emphasis in original). In a gloss on the biblical notion that each species "produced after his kind," she sought scriptural affirmation of "the fundamental law that 'like begets like,' the law of hereditary transmission." The author construed God's breathing life into the clay and thus producing the first man as "endowing man with the divine mind." [32]

Within her implicit theological anthropology, Teats construed the body as loathsome. Building on the biblical phrase, "God is spirit," she located the image of God in the human spirit, an entity she believed to be distinct from yet connected with the mind. "In the state man occupied before the fall," she wrote, "the spirit was to occupy the throne, so to speak, mind was to draw from the spirit, and the spirit and mind were to control and direct the body." The fall reversed the order, so that the body became "enthroned," and the curse of the human race is its thralldom to the body. "The lesser undertook to rule over the greater (body over spirit). . . ." [33]

Teats's interpretation of the fall followed a prominent tradition that interpreted the Genesis narrative in sexual terms. Though providing no detailed account of this hermeneutic, Teats drew upon personal testimony, an authoritative evangelical metier. "As the years passed on, and I had wider opportunities of studying humanity," she submitted, "it was borne in upon my soul that the sin through which the race fell (and for that matter, is still falling) was a sexual sin; the perversion of the power and object of parentage." As textual backing for this claim, Teats referred to the postlapsarian shame of Adam and Eve over their nakedness, and their efforts to conceal it. [34] This theory linking sex and original sin had deep roots in the Augustinian tradition, but would prove a difficult sell in modern times.

For Teats, this dire view of human sexuality had far-reaching social ramifications. In phrases bearing significant parallels to the starkly elitist rhetoric of activists in the eugenical sterilization movement, she opined that:

> The great and rapidly increasing army of idiots, insane, imbeciles, blind, deaf-mutes, epileptics, paralytics, the murderers, thieves, drunkards and moral perverts are very poor material with which to "subdue the world," and usher in the glad day

32. Teats, *Way of God in Marriage*, 9–11.

33. Ibid., 17.

34. Ibid., 23–26.

> when "all shall know the Lord, whom to know aright is life ever-
> lasting." There are hundreds and thousands of men and women
> to-day (sic) to whom in the interests of future generations, some
> rigid law should say, "Write this one childless." Men and women
> whose habits of life are such as to curse their offspring, should
> be prohibited from marrying.[35]

In a later section, she connected such unfortunates with Malachi's prophetic
rebuke of postexilic Israel's offering of blind, lame and sick animals as sac-
rifices. She scoffed at the notion that "the lame, halt, deaf, blind, mutes,
imbeciles, idiots, drunkards and moral perverts" could be properly called
"God-given children," or considered a proper offering and gift to God.[36]

In prohibiting this laundry list of "defectives," from procreating, Teats
was a harbinger of the involuntary sterilization favored by the leaders of
the American eugenics movement. Her use of eschatological language drew
upon long-standing themes in American Protestantism concerning the
"Kingdom of God." Such rhetoric prepared devout Protestants to accept the
doctrines, and by 1907 state-level involuntary sterilization policies, of the
eugenics movement.[37]

What were the specific "habits of life" to which the Teats was referring?
The short answer is "marital incontinence" and "the use of strong drink."
The long answer is more complex. The long answer involves exegeting the
putative effects of such factors as the mental and spiritual states of the parents at the moment of a child's conception and especially the mental states
of the mother during fetal gestation. In such a view, not only physiological
but also mental states of the pregnant mother could dramatically influence
the physical, mental, and moral development of the child, as well as his or
her likely future character.

Such claims are only understandable against the backdrop of a neo-
Lamarckian understanding of heredity dominant at the turn of the century,
inherited from Edward Drinker Cope and his followers, which in turn
partook of a much longer tradition of viewing heredity as highly plastic.
Such claims evinced an extremely soft view of heredity while the fetus was
still in the womb. Yet, paradoxically, Mary Teats could also espouse a very
deterministic view of the persistence of such influence during the lifetime of
the affected offspring once they were born. It is also possible that Teats was
influenced by Memory Theory, espoused by Germans Hering and Semon,
and American Samuel Butler. In the 1920s J. Parton Milum described this

---

35. Ibid., 30.
36. Ibid., 43–44.
37. See Dowbiggin, *Keeping America Sane*, vii-viii.

theory thus: "as habit may modify the form of the individual, so instinct may have modified the species." This theory gave a heavy role to the unconscious as "responsible for deeds and processes." Samuel Butler went so far as to assert that heredity is memory, and represented an attempt to biologize the mental processes and discern their intergenerational effects.[38] This fit well with certain forms of popular science that held the mental life of the mother had profound physiological effects on the formation of habits across the stages of life, from fetus to child and even to adult.

For biblical verification of her views on the dictates of human heredity, Teats looked to the example of King David. David's sins, she declared, were indicative of "the characteristic of sensualism that had been handed down through the long line of hereditary taint . . . ." She even proclaimed that: "If David's parents had had the privilege which all Christians have to-day, of understanding the laws of procreation and prenatal influence . . . it is doubtful if David would have succumbed to the temptation." She went on to cite Psalm 51:6, in which David, wracked with guilt for his dalliance with Bathsheba and subsequent arrangement of her husband's death, wrote: "Behold, I was shapen in iniquity and in sin did my mother conceive me." Here she commented: "It is possible that he felt that during his prenatal life, his parents had violated God's laws of continence, resulting in his life even in the embryonic state, being mortgaged repeatedly to a life-struggle against sensual desires."[39] Later she lamented that "David bequeathed a legacy of sin and shame to his children which caused the brother to seduce the sister, and her ruin to be avenged by a brother staining his hands with that brother's blood." For Teats, Solomon's propensity to multiply wives and concubines was a further evidence of a parentally predetermined sensualism.

The hereditary determinism of such passages is striking, particularly given the malleability predicated of the developing fetus in the womb. Though responsible for her own interpretations, as well as her framing of the scientific issues, Mary Teats was at least in part reflecting a state of confusion over heredity prevalent at the time among professional scientists who studied it. At the same time, the growing prestige of science in its increasingly professionalized and specialized development was eagerly seized upon by many interest groups in the era, certainly not just the WCTU.

The level of authority attributed to science even by very religious persons stands out in *The Way of God in Marriage*. Ms. Teats, in her chapter "Science of Life," boldly asserted: "Had the people of the church of Christ coupled the science of life all these past centuries with the Apostolic creed,

38. Milum, *Evolution and the Spirit of Man*, 63–64.
39. Teats, *Way of God in Marriage*, 34–35.

giving the divine law of the conservation of vital force its rightful recognition, Christians would not today be producing sinners."[40] Teats offered a strongly physiological interpretation of the perfectionist impulse in Protestant social reform. In her view, the practice of marital continence was so firmly established, by scripture and by science, that it held the status of dogma. This was a perfectionism that stressed biology to such an extent that other components of the human personality became obscured and subordinated.

Mary E. Teats was not alone in eagerly seizing on one strand or another of the developing science and building upon it a *cause celebre* complete with government legislation. "Eugenics, with or without the name," writes historian of science Philip J. Pauly, "was a significant bandwagon well under way in the United States when academic biologists first joined it in the years around 1910." The main impetus for these scientists' support for eugenics was not simply the growth of the field of genetics but rather as a component of "their heightened interest in social problems at the climax of the Progressive Era."[41]

Mary Teats and other WCTU members saw eugenics as one of many means to the end of social reform and social purification under the tutelage of middle-class Protestant women. As a means of purifying America and ushering in Christ's kingdom on the wings of Protestant reform, eugenics held an appeal both religiously sanctified and patriotically pragmatic.

Blanche Eames and Practical Eugenics

Another contribution to the marriage manual genre, was *Principles of Eugenics: A Practical Treatise* by Blanche Eames (1870–1962). In her preface the author extolled the "few progressive states" and numerous prominent cities which had formed "Societies of Social Hygiene" to disseminate eugenics principles to the nation's youths. Eames singled out for approbation such organizations as the Eugenics Education Society of London, the Eugenic Section of the American Genetic Association, and the above-treated Correspondence School of Gospel and Scientific Eugenics of Mary E. Teats. Eames offered chapters on reproduction, heredity, "race poisons," instruction in sex truths, tobacco and drugs, prenatal nutrition and environment, and continence.[42]

---

40. Ibid., 74.

41. Pauly, *Biologists and the Promise of American Life*, 216.

42. Eames, *Principles of Eugenics*, 7–12.

Citing prominent eugenicist H. H. Goddard of the Vineland, New Jersey Institution for the Feeble-minded, and the American Genetic Association, she stated: "The fact that feeble-mindedness, epilepsy, insanity and other forms of degeneracy are passed on from ancestor to descendants, is now well authenticated by careful and systematic investigations." Decrying the annual expense of caring for such persons, she posed the rhetorical question, "Would it not be the part of discretion, and one might even say of true patriotism, to absolutely prohibit these not-up-to-the-standards from raising families?" Applauding those states that had already passed involuntary sterilization laws, Eames called for education, sterilization, and segregation as necessary means to the goal of a healthy society. Acknowledging critics who regarded sterilization as both cruel and infringing upon individual liberties, Eames wrote, ". . . it must be remembered that the community, the State and the country also have rights that can justly be pitted against those of the individual." Such rhetoric betrays a crisis mentality that saw hereditary impurity or imperfection as a threat to be met with drastic measures.[43]

In the chapter entitled "Race Poisons," Eames drew the connection between alcohol —the most dastardly nemesis of the WCTU— and heredity. Citing prominent British prohibitionist physician Caleb Saleeby, she declared that alcohol "acts as a poison to the germ cell." Yet physiological and environmental factors worked together to blight the home of the alcohol abuser. Eames cited a study by T. Alexander MacNicholl, a scientist favored by Theodore Roosevelt, concerning children of drinkers and non-drinkers. Pegging a laundry list of physical maladies to alcohol abuse, Eames cited MacNicholls' study as showing that "In the children of total abstainers, 90 per cent were normal in mind and body as against 7 per cent of the drinkers' children."[44] As to other possible environmental factors leading to blighted homes, such as poverty, Eames averred that alcohol was the main culprit for these as well. Mental derangement, crime, prostitution, syphilis, gonorrhea, and sterility could all be traced back to alcohol as efficient cause.[45]

In a passage on sex education, Eames averred that young men ought to be instructed in "the beneficial effects of continence and the injurious consequences of squandering life force." In a later chapter on continence, Eames opined that such "squandering" yielded languid spermatazoa, lacking in vitality and vigor. Eames blamed the solitary vice for wide-ranging social ills, declaring that "this may be the chief reason why the civilized

43. Ibid., 27–31. On Goddard, Leila Zenderland in particular takes into account his evangelical Quaker upbringing and its lifelong effects on his thought and ethos. See Zenderland, *Measuring Minds*, 16–43; 363–64.

44. Eames, *Principles of Eugenics*, 35.

45. Ibid., 36–42.

human race is cursed with so many weaklings and degenerates, and why few really are strong and healthy."[46] Both males and females needed to "study the laws of heredity so as to understand that their thoughts and deeds of to-day will have an influence upon their posterity of to-morrow."[47]

When addressing the issue of prenatal nutrition and environment, Eames demonstrated a basic grasp of the major figures in the debates over hard and soft heredity current in her day. At the same time, she cast her lot in with those who favored a soft interpretation of heredity. She claimed that both Spencer and Darwin acknowledged that some modification of species occurs through "the inherited effects of the use and disuse of parts." She also cited David Starr Jordan and Luther Burbank, prominent eugenics figures, as favoring the inheritance of acquired characteristics.[48] She challenged the view of "Weismann and his followers" that acquired characters are not heritable. Drawing on a common-sense interpretation of the variation evident in siblings of shared parentage, Eames panned the notion that germ plasm was immutable. Making a point very similar to that of Mary E. Teats, she countered the theorizing of Weismann's students:

> And when they say that these differences are caused by variations in the germ plasm, it may be asked if these variations are not caused by the influence of the parents' surroundings and thought-life upon their reproductive cells. Thought is the mightiest thing in the Universe and may well it be believed that it is sufficiently powerful to take hold of an individual's whole being, his life-giving faculties included; the differences in character between members of the same family being due to different thoughts, purposes and ambitions existing in the minds of the parents before the conception of these children.[49]

Later, Eames developed this line of argument from the waning nineteenth-century idealist tradition even further. The child derives its nourishment from the blood of the mother. The mother's thoughts, if they are negative, contain "poisons" that enter into the bloodstream and damage the fetus. Positive thoughts, however, have the opposite effect: "But on the other hand, if she is filled with hope, peace and love, then secretions which are

---

46. Ibid., 85.

47. Ibid., 47.

48. "Acquired characters *are* transmitted," Burbank would insist, and "*all* characters which *are* transmitted have been acquired" by repetition and "become inherent, inbred, or 'fixed,' as we call it." Burbank, *Training of the Human Plant*, 82.

49. Eames, *Principles of Eugenics*, 68–69.

beneficial to the system are thrown into her blood and will help to produce the normal development of the child."[50]

Eames, much like Mary Teats, wove together scientific, historical, and theological themes in a manner calculated to optimize the authority of her claims. Much like the revivalist sermon of the frontier preacher, or the jeremiad of the Puritan divine, she lashed out at perceived vice with evangelical zeal. Using the growing prestige of science, and particularly the conclusions of the emerging science of heredity, she could construct a thoroughly modern and sophisticated case. Playing upon themes of patriotism and progress, she could tap into the fervor of nationalism and the rhetoric of Christian civilization in the early years of World War I. Included was a strong strain of sarcasm toward the current "civilization" in contrast with the real civilization that social reformers were striving toward. All these themes coalesced as her book reached its end:

> When we consider that for so many centuries God's law of the conservation of life forces has been most flagrantly disobeyed, we understand why humanity is so seamed and scarred to-day (sic). But it need not remain in that condition. Through His laws of reproduction; natural selection, heredity, including the transmission of both inherent and acquired characteristics; and gestative influence, God has most wonderfully provided for the development and ultimate perfection of mankind. . . . It is only perverted nature and the abnormal habits of so-called civilization that are in the way of the fulfilment of such a hope for the race. . . . Now, in the fullness of time, and for some good reason, we are treated as children no more, and the conscious fashioning of the human race is given into our hands. Let us put away childish things, stand up with open eyes, and face our responsibilities.[51]

Physicians and Warnings about Hereditary Inebriety

The rise of scientific temperance would have been deeply hampered had it not been for the involvement of prestigious physicians devoted to prohibition. Among these was J. M. French, M.D., medical director of the Elmwood Sanitarium. In the *Quarterly Journal of Inebriety* of the American Association for the Study and Cure of Inebriety, French wrote the following

50. Ibid., 76.
51. Ibid., 89–90.

warning in January, 1898. The essay, entitled "The Prognosis of Inebriety," indicated the shift in treatment that was well underway from a moralistic to a medical model of intervention. Still, this shift was a slow one wherein moral and religious categories held a central role in assessing the impact of alcohol on families. Thus, regarding the heredity of the problem, French urged some knowledge of a given inebriate patient's physiology and habits of life. Included among these were "His mental ability and moral character" as proper objects of study. French went on to query: "Was he born sound. . . or is he by nature a degenerate—'a morbid deviation from a normal type?' Is he inclined by heredity, either direct or indirect, to inebriety, narcomania, epilepsy, insanity, or any other form of nervous instability?" French went on to make a far-reaching claim that drew upon Morel's degeneration theory as well as on the Bible:

> In studying heredity, which is the first of the remote or predis-
> posing causes of relapse to be considered, it does not so much
> matter whether it is direct or indirect, immediate or remote. In
> either case, the fact remains that a large proportion, probably
> more than half of the whole number of inebriates, were born
> wrong, endowed by nature with a defective nervous system, and
> are only to be cured by "beginning with their grandfathers." It
> may be that there was inebriety in their ancestry, or just as likely
> it may have been some other form of degeneracy and disease
> nervous or constitutional.[52]

Some brief commentary on this passage is in order. First of all, French assumed that past causes were predisposing the alcoholic to relapse. These causes were further to be hierarchically arranged, and heredity was the first or top cause of relapse. Secondly, "probably more than half of inebriates were born wrong" was shoddy science at best even at the time, and statistically vague. This sort of heavily value-laden and anecdotal rhetoric was endemic to the eugenics mindset. Thirdly, even if it wasn't inebriety, some other form of ancestral degeneracy had rendered the client's inebriate client's state highly likely in French's opinion. Thus degeneracy was a flexible construct capable of diagnosing a wide range of maladies and behaviors. Fourthly, the notion of needing to reach the grandfather to treat the grandson's inebriety is an allusion to the degeneration theorists' frequent gloss on God's punish-ing the sin of the fathers on the children "to the third and fourth generation" from the book of Exodus in the Old Testament. Thus biblical allusion and moralistic rhetoric and blended with degeneration-science-as-heredity in ways that would soon become much more clearly anachronistic.

52. French, "The Prognosis of Inebriety," 19–21.

Conclusion: Temperance and Popular Eugenics

If these figures are any indication, the WCTU writings and other sources of literature of social reform within the ambit of temperance agitation proves to be a trove of eugenics thought.[53] Much of the historiography on the eugenics movement has focused on males—not surprisingly, since its most prominent and prolific promoters were men. Yet the widespread popularity of eugenics, and the ability of eugenics organizations to promote their ideas at all levels of society, in rural as well as urban settings, presupposes a broad base of grass-roots support, or at least non-resistance.[54] The WCTU and other prohibition organizations were able to facilitate a constitutional amendment, restricting an activity as pervasive and popular as imbibing alcohol. Therefore, it is likely that other public purity endeavors of the WCTU garnered broad support from Protestants across the ideological and theological spectrum.[55] The next chapter represents a shift of focus to the quintessentially problematic American issue of race, and an exploration of the degeneration theme as it relates to scientific racism and both its religious promoters and critics.

53. In a chapter entitled "Alcohol and Eugenics," Edith Smith Davis, Superintendent of the WCTU Scientific Temperance Department, declared: "That there is nothing new under the sun receives confirmation in the fact that the law of Moses is the law of Eugenics—that the sins of the fathers shall be visited upon the children unto the third and fourth generation. Likewise the children shall have health and happiness whose parents have lived according to the law of life which is the law of God." Davis, *Compendium of Temperance Truth*, 116.

54. Diane Paul has noted the rampant prevalence of eugenics promotion in high schools and at county fairs across America prior to WWII. Paul, *Controlling Human Heredity*, 10–21.

55. Zenderland offers many examples of figures who melded eugenic ideals with biblical proof-texts thus to: "illustrate efforts by American Protestants to reconcile age-old Christian messages with new eugenic doctrines. In doing so, their writings blurred together the many meanings of a good inheritance—popular, biblical, and now biological." Zenderland, "Biblical Biology," 522–23.

CHAPTER 9

Degeneration and the Race Question

I believe that all men, black and brown and white, are brothers, varying through time and opportunity, in form and gift and feature, but differing in no essential particular, and alike in soul and the possibility of infinite development.

—W. E. B. DuBois, 1904[1]

Introduction

THE ENGLISH TRANSLATION OF Arthur de Gobineau's *Essay on the Inequality of the Human Races* (originally published 1853–1855) only appeared on the American scene in 1915, but its ideology of degeneration had already become diffuse in transatlantic thought over several decades. By that time fears of the spread of the European war, along with intensified nationalistic ideologies and American elites' fears of the rapid immigration of eastern and southern Europeans of allegedly "inferior stock" were proliferating. De Gobineau's work fit hand in glove with such fears and the crafting of social policies to restrict the immigration and procreation of those deemed "unfit" by nativists. He declared:

> The word *degenerate*, when applied to a people, means (as it ought to mean) that the people has no longer the same intrinsic value as it had before, because it has no longer the same blood in its veins, continual adulterations having gradually affected the quality of that blood. In other words, though the nation bears the name given by its founders, the name no longer connotes the same race; in fact, the man of a decadent time, the *degenerate*

1. Hamilton, *Writings of W. E. B. DuBois*, 56.

man properly so called, is a different being, from the racial point of view, from the heroes of the great ages.[2]

The father of the eugenics movement, Francis Galton, believed humans had an influence upon the racial character of humanity over time. Galton had unapologetically applied the core insight of Charles Darwin, namely the theory of natural selection, to the races of humanity:

> There exists a sentiment, for the most part quite unreasonable, against the gradual extinction of an inferior race. It rests on some confusion between the race and the individual, as if the destruction of a race was equivalent to the destruction of a large number of men. It is nothing of the kind when the process of extinction works silently and slowly through the earlier marriage of members of the superior race, through their greater vitality under equal stress, through their better chances of getting a livelihood, or through their prepotency in mixed marriages. That the members of an inferior class should dislike being elbowed out of the way is another matter; but it may be somewhat brutally argued that whenever two individuals struggle for a single space, one must yield, and that there will be no more unhappiness on the whole, if the inferior yield to the superior than conversely, whereas the world will be permanently enriched by the success of the superior.[3]

While admitting that the extinction of some races would happen gradually, Galton was clearly not averse to guiding: "the power in man of varying the future human stock" by conscious intervention. For Galton, such a power "vests a great responsibility in the hands of each fresh generation" which, by his lights, in 1908 had not adequately been used.[4]

Eugenicists in Europe and America were quite willing to advance various social policies to speed up the process of winnowing out so-called inferior races. On the question of race very disparate and contradictory positions were taken within the broad arena of Protestant social reform, but prejudicial approaches tended to gain the upper hand amid the ethos of eugenics.

---

2. Gobineau, *Inequality of Human Races*, 25.

3. Galton, *Inquiries*, 200–201; cf. Redfield, *Control of Heredity*, 194–205, chapter 12, entitled: "Races of Men," where he mirrored Galton's analysis of race characteristics and arrangement of races in a hierarchy of fitness.

4. Galton, *Inquiries*, 206.

Debates over the Biology of Race

Discussions of race in the late nineteenth and early twentieth century may be framed within a history of the relationship between nature and nurture, biology and culture. Whereas the nineteenth century bequeathed to the twentieth a strongly biological hierarchy of the races, by the late 1920s the role of culture in an intellectual approach to race began to become much more pronounced. Historian Colin Kidd has ably surveyed the rise of the notion of Aryan supremacy among European and American Protestant elites based on phrenology and comparative philology in the 1800s. This attitude was pervasive and far-reaching.[5] During the second half of the twentieth century a sea change has occurred through research into human biology and genetics. These investigations have yielded a strong consensus in favor of the view that race is not a coherent biological concept. Racial variation in biological terms is found mainly in relatively trivial matters such as levels of melanin in skin cells and variations in hair texture. The genetic differences between persons of differing skin colors are extremely small, and other points of variability are much more genetically revealing. In a discussion of race and biology, a major work in the field concludes:

> There have never been any biological races, any distinct subdivision of our species, and there are none now. All modern human populations have the same reproductive capacities, the same intellectual capacities, the same emotional, behavioral, and moral capacities.[6]

Further, the possibility of blood transfusion across the vast spectrum of the human species bespeaks the fundamental unity of humanity from a biological perspective. Despite criticisms by scientists, the Red Cross bowed to political pressure and in 1941 announced its segregation of the blood supply during World War 2.[7] Furthermore, claims that race is strongly linked to IQ, widely favored by eugenicists and immigration restriction enthusiasts, have not stood up to sustained and rigorous scrutiny.[8]

One standard strand of evidence of the strong intellectual acumen of African American humanists was the blossoming of the Harlem Renaissance, first enunciated in a prominent article by black philosopher Alain Locke in 1925. This oft-noted rise of the phenomenon of powerful black voices in the arts and literature was centered in Harlem, New

5. Kidd, *Forging of Races*, 168–202.

6. Mukhopadhyay, et al., *How Real is Race?*, 83.

7. David Livingstone Smith, *Less than Human*, 182.

8. Montagu, *Race and IQ*, 190–206, 230–47.

York. It sparked a debate amongst black intellectuals as to whether racial pride should be a point of emphasis in the project of uplift and self-improvement. The intellectual rigors of this internal 1920s debate inherently demonstrate the absurdity of eugenic claims at the time of invidious links between race and IQ.[9]

Race as a cultural phenomenon differs from race as a biological category. The reality of race as a socio-cultural discussion seems especially endemic to North American culture, spawning a vast literature far beyond the scope of this treatise.[10] This chapter is designed to introduce a descriptive account of the nexus between eugenic thought, biology, and religion in the period prior to the Second World War. Debates between those who espoused hierarchical arrangement of the races and those who argued for an anti-hierarchical construction thereof cut across Protestant thought. Articulated fears of race degeneration represent a typical iteration of such thinking. Though both whites and blacks were experiencing biological degeneration according to some prominent eugenicists, they often asserted that African-Americans were in the greater and more immediate danger because of their ostensibly inferior biological starting point.[11] African-American scholars as well as some Caucasians pushed back against such prejudicial thinking, and allied themselves with more egalitarian strands of thought emerging variously from theology and the rising new field of cultural anthropology.

The issue of race is much broader than the black/white dichotomy that has historically prevailed in the American press. In the late 1800s elites even expressed fears over certain regions of Europe sending too many of their people to American shores via a rapidly growing number of eastern and southern European immigrants. These elite fears took popular literary form, and perhaps none more popular than the writings of Protestant apologist Josiah Strong. His writings gave voice to racial and religious fears in the age of expanding immigration.

9. Holloway, "Harlem Renaissance Scholars Debate the Route to Racial Progress," 60–63; Lecklider, *Inventing the Egghead*, 92–115; and Gates, "Harlem on Our Minds," 1–12.

10. An exploration of the voluminous writings of Cornel West, Henry Louis Gates, James Cone, Thomas Sowell, Wilson Jeremiah Moses, and William Julius Wilson would be the recommended point of entry into these debates.

11. On heredity and race during this period see Cravens, *Triumph of Evolution*, 157–90.

Fears of Immigrants and the
Changing Urban Landscape

A Congregationalist minister and successful essayist, Josiah Strong (1847–1916) observed the rising tide of immigrants and expressed fears common among old-stock Anglo-Americans. This expression took the form of a popular tome entitled *Our Country*, published in 1885. In a typical passage, Strong offered the following warning:

> Immigration brings unquestioned benefits, but these do not concern our argument. It complicates almost every home missionary problem and furnishes the soil which feeds the life of several of the most noxious growths of our civilization. I have, therefore, dwelt at some length upon its future that we may the more accurately measure the dangers which threaten us.[12]

The bulk of Strong's warning was devoted to apprising his readers of several "perils" resulting from the increase and the overall quality of the immigrants arriving daily on American shores. These ostensible perils, each receiving a chapter-length treatment, were: immigration, Romanism, Mormonism, intemperance, socialism, wealth, and the city.[13]

Regarding urbanization, Strong lamented the rise of cities teeming with immigrants resistant to assimilation. He noted with trepidation that in 1790 only one out of thirty residents of the U. S. lived in a city larger than 8,000; whereas by 1880 some 22.5 percent of the populace lived in cities thus defined. Strong described the city as "a serious menace to our civilization," wherein "Romanism" found its greatest strength. The saloon, intemperance, and "the liquor power" also proliferated in the city. Wealth and ostentatious displays thereof were an evil concentrated in cities, and at the other end of the spectrum cities were congested and overcrowded with tenement slums. Such rhetoric indicates a belief that the middle class was the moral center of societal gravity, and the target audience of the book. Socialism took root and grew in its baleful influence in cities. Political corruption under the boss system was the characteristic feature of this degenerating form of city government. Meanwhile churches of the evangelical variety favored by

---

12. Strong, *Our Country*, 40.

13. Religious historian Robert Handy points out that: "After 1880 . . . many Protestants found themselves increasingly alarmed by the waves of so many newcomers from conspicuously different backgrounds, outsiders who seemed to jeopardize their social and religious hegemony in the culture." Handy, *Undermined Establishment*, 17.

Strong, and which could serve as an ameliorating and civilizing influence in the city, were simply swamped and thus unable to keep up with their task.[14]

Strong's solution to these problems was the promotion of home missions, a common refrain among old-stock evangelicals of the time. He promoted benevolence work, and sacrificial giving to charity, pleading that "the gospel is the radical cure of the world's great evils."[15] For Strong there was no distinction between Christianization and the making of a solid citizen. "Christianize the immigrant," he preached, "and he will be easily Americanized."[16]

Science and Immigration

Religious and missionary rhetoric was not the only medium for assessing immigrants. Antipathy toward immigrants often took on a scientific cast among elites at the end of the nineteenth century. Eminent historian of the immigrant experience, Oscar Handlin, has pointed out the uses made of science by anti-immigrant nativists. "Anthropologists measured thousands of skulls to arrive at a precise index of race"; he observed, "and geneticists showed statistically that no degenerate or feeble stock could ever be made healthy by education or favorable environment." The conclusion for white Americans convinced of the hierarchy of races, was that "superior and inferior races could not live together on any other terms than as masters and slaves."[17] Immigration restriction enthusiast Lothrop Stoddard enunciated the pessimistic position in 1920 that "the introduction of even a small group of prolific and adaptable but racially undesirable aliens may result in their subsequent prodigious multiplication, thereby either replacing better native stocks or degrading these by the injection of inferior blood."[18]

John Miley on the Unity of the Races

Not all late nineteenth-century Protestant intellectuals agreed with such a biased stratification of the human family. Drew University theologian John Miley (1813–1895) insisted on the oneness of the races in humanity (monogenism). His *Systematic Theology* made the case that "the distinctions

14. Strong, *Our Country*, 128–36.
15. Ibid., 209.
16. Ibid., 210.
17. Handlin, *Americans*, 304.
18. Stoddard, *Rising Tide of Color Against White World Supremacy*, 252.

are superficial, and the result of local influences."[19] Miley regarded humans
as one in physical characteristics of the human body's chemistry, anatomy,
physiology, and pathology (susceptibility to disease or injury). He further
urged the unity of humanity in its psychology. Any mental differences between races were superficial, while they held "a oneness in all the intrinsic
facts of mind." In terms of sensibility, adaptability, intellect, rationality, morality, religiosity, the races were alike. In stark contrast to the racialized rhetoric of scientific racism, Miley admitted that while the moral and religious
nature "may sink to barbarism and idolatry in the Caucasian," it contrast-
ingly "may rise to the highest moral and Christian life in the Mongolian and
Negro.[20] He cited natural historians and ethnologists such as Quatrefages
and Prichard to observe universal endowments of mental capacity and that
"the same inward and mental nature is to be recognized in all the races of
men."[21] Further, citing the "law of hybridity which limits the production of
a permanently fruitful progeny to the species" as "one of the most obvious
laws of natural history" Miley strengthened his case for the essential unity
of the races capable of interbreeding.[22]

Miley also cited the lineal connections between the human languages
to argue for the unity of humanity. Using the field of comparative philology, he argued that "peoples widely separated in place, and representing
very distinct racial types, were originally one family and one blood."[23] This
position stood in stolid opposition to polygenism, the view that the various
races of humanity had separate, and thus unequal, hereditary origins. While
Miley showed an openness to the polygenism of prominent figures like Harvard's Louis Agassiz, a study of the fruits of ethnology and natural history
led Miley to conclude that: "The results already attained render groundless
the distinctions of race for a plurality of origins, and prove beyond question that more or less of the several species as held by polygenists are mere
varieties of the one species."[24]

A common trope among monogenists, and especially abolitionists of
the Civil War and Reconstruction eras, was the citation of a quotation from
St. Paul: "And [God] hath made of one blood all nations of men for to dwell
on all the face of the earth . . . ."[25] Miley nuanced the interpretation of this

19. Miley, *Systematic Theology,* 1:378.

20. Ibid.

21. Ibid., 1:379.

22. Ibid., 1:379–82.

23. Ibid., 1:383.

24. Ibid.

25. Acts 17:26, King James Version.

passage to go beyond the unity of the human family at the species level, to urge the "genealogical oneness" of the human race descended from Adam. In fact, for Miley, "the common sinfulness of the race would not in the deep sense of Paul be consequent with the sin of Adam without a common genealogical connection with him."[26]

Miley forcefully attacked a popular nineteenth-century polygenist theological theory known as pre-Adamism.[27] Polygenists ridiculed the story of Adam and Eve because the propagation of the human race required brothers and sisters to mate. In reciprocal response, Miley mocked the implication of polygenism that "only the coincident evolution of two human beings, respectively male and female, could meet the lowest requirement for the inception of a human race." He further postulated that the offspring of such an unlikely pairing would be brothers and sisters, such that the consistent polygenist would need two couples to evolve simultaneously "so as to provide for marriage without the consanguinity of brother and sister."[28] Doubling down on this theological bet, Miley also pointed out that under pre-Adamism different races represented separate origins of the races (Caucasian, Negroid and Mongoloid), yet to avoid the consanguinity problem, each *race* would require the independent and simultaneous evolution of two pairs. If the races were not fully independent in origins but branched from one another, the problem of race-mixing and diminishment of pure stock arose, unless (again) a simultaneous male and female evolution of a *de novo* race had occurred.[29] By such logic Miley subjected pre-Adamism to a trenchant critique based on diminishing probabilities and racial dilution at the earliest stages of development. As we shall now see, however, other religious figures used theological and biblical tropes to reinforce rather than to challenge common racial prejudices of the time.

Religious and Scientific Racism

The nineteenth-century literature treating science, religion, and race represents a tragic chapter in the interplay of potent cultural forces granting staying power to scientific racism. This literature developed out of a convergence of the rise of phrenology, heated political debates over the abolition of slavery, and theological disputation between monogenists and polygenists. The science of phrenology in the mid-nineteenth century buttressed racist

26. Miley, *Systematic Theology,* 1:384–85.

27. Ibid., 1:385–87; cf. Livingstone, "Origin and Unity of the Human Race," 452–57.

28. Miley, 1:387.

29. Ibid., 1:388–89.

claims for the inherent inferiority of those of African descent. Theological exponents of "the mark of Cain" and "the curse of Ham" in the early chapters of Genesis gave such stories a virulently prejudicial twist in their treatises on biblical themes.

Racialist ministers were heavily engaged in writing novels and essays at the turn of the century, some of which treated the theme of race as biologically determined and stratified into superior and inferior manifestations. Thomas Dixon, a Baptist minister whose greater fame was as a novelist and movie producer, was the most notorious. Dixon's books *The Leopard's Spots: A Romance of the White Man's Burden* (1903) and *The Clansman: A Historical Romance of the Ku Klux Klan* (1905) clearly espoused the inherent inferiority of African-Americans along with white supremacy. Dixon was also the mastermind behind the infamous pro-Klan film, *Birth of a Nation*.[30] Dixon's novels spawned many books and essays by African-Americans who articulated arguments to counter the notion of their inherent racial inferiority. J. W. Grant's *Out of Darkness* (1909), Sutton E. Griggs, *The Hindered Hand* (1905), Thomas H. B. Walker's *J. Johnson or the Unknown Man* (1915), and Herman Dreer's *The Immediate Jewel of His Soul* (1919) are a few prominent examples of this response identified by Dixon's biographer.[31]

Some racialist authors sought to construct even more explicitly biblical arguments for the inherent inferiority of blacks. Some prominent examples included D. G. Phillips, *Nachash: What Is It? Or An Answer to the Question, "Who and What is the Negro? Drawn from Revelation* (1868); A. Hoyle Lester, *The Pre-Adamite, or Who Tempted Eve? Scripture and Science in Unison as Respects the Antiquity of Man* (1875); and Charles Carroll, *The Tempter of Eve; or, The Criminality of Man's Social Political, and Religious Equality with the Negro, and the Amalgamation to Which These Crimes Inevitably Lead: Discussed in the Light of the Scriptures, the Sciences, Profane History, Tradition, and the Testimony of the Monuments* (1902).[32] These works constructed African-Americans as "beasts," and Carroll even went so far as to interpret the serpent in the Genesis account of the Garden of Eden as "a negress."[33]

---

30. Survey Thomas Dixon, *Leopard's Spots*; Dixon, *The Clansman: An Historical Romance of the Ku Klux Klan*; and on Dixon's film efforts, see Raymond A. Cook, *Thomas Dixon*, 109–122.

31. Cook, *Thomas Dixon*, 133–38.

32. All these are reprinted in their entirety in John David Smith, *Anti-Black Thought 1863–1925: "The Negro Problem,"* especially Vol. 6, *The Biblical and "Scientific" Defense of Slavery.*

33. See a description of such works, and a detailed analysis of Carroll's *Tempter of Eve*, in Stokes, *Color of Sex*, 82–107.

James M. Boddy and the Critique
of Racist Phrenology

Formulating a cogent response to such a barrage of literature in support of racial inequality during this era required the talents of persons trained in both theology and science. One such individual was James M. Boddy (1886-?), whose varied and impressive career has received little attention in the secondary literature.[34] Boddy, an African-American physician and Presbyterian theologian, grew up during the Reconstruction Era, having been born in Wrightsville, Pennsylvania in 1866. He received his A. B. at Lincoln University, a Presbyterian college for African-Americans, in 1890. After graduating with a Master's Degree from Princeton Theological Seminary in 1895, Boddy was ordained into the Presbyterian ministry, and he pastored the Siloam Presbyterian Church in Elizabeth, New Jersey. He pursued a medical education after seminary, and received the M.D. in 1906 from Albany Medical College in New York. Boddy served as minister to numerous churches across the U.S., including Troy, New York; Little Rock, Arkansas; and St. Paul, Minnesota.[35]

As one religious historian has noted, during this period many historians of race were ordained ministers.[36] African Americans writing in this genre set forth a positive and progressive outlook for the black community, in contrast to more invidious accounts of past and future set forth by religious and scientific racists. James M. Boddy was one important voice in this effort.

Boddy espoused ideas that can be bewildering for the later reader unfamiliar with the state of scientific conjectures on race at the turn of the century. His essay "The Ethnology of the Japanese Race" argued that "the physical characteristics of the Japanese race, sciences traces to the Oceanic Negro." In other words, Polynesia functioned as an ethnic link between the Japanese and Africans, thus the Japanese were "part and parcel of the Negro race." This material Boddy did not merely invent, but drew from works in the "science of ethnology," many of which could be characterized as nineteenth-century race histories.[37]

More relevant to a discussion of "race degeneration" in the West was Boddy's 1905 essay "Brain Weight and Intellectual Development: Physical Variations of the Negro and the Anglo-Saxon Races." In this essay he melded

34. I am indebted to Quinton Dixie of the Religion Department at the University of Indiana for making me aware of James M. Boddy.

35. *Who's Who of the Colored Race*, s.v. "Boddy, James Marmaduke."

36. Maffly-Kipp, "Mapping the World, Mapping the Race," 610–26.

37. Boddy, "The Ethnology of the Japanese Race," 577–85.

his training in science and theology to form an argument against those who promoted white supremacy under the guise of phrenological science. His method of critique, however, was oblique, in that he refrained from mounting a direct challenge to phrenology, lest perhaps he fall prey to the charge of being unscientific in his views. Thus Boddy in some respects contributed to the perpetuation of the categories of "the science of race." But from within that paradigm, he sought to create space for an optimistic assessment of the prospects of African-Americans for future success. This approach stood as a challenge to the widespread prejudicial application of degeneration theory to non-white races.

Boddy accepted the premise of "the savants" (i.e., scientists) that measurements of skulls derived from both Negroes[38] and Anglo-Saxons indicate a difference in brain weight between the races. Yet he claimed these differences were not the original will of the Creator of humanity, but "accidental and acquired variations," caused by natural phenomena "such as the solar heat of the tropical zone," and other environmental conditions like atmospheric and dietary conditions.[39]

Boddy built upon the standard descriptions offered by phrenologists, i.e., those who measured the features of the skull as a means of scientifically predicting intellectual ability for persons of various races. Boddy parted ways with many phrenologists about the conclusions they reached concerning the impact of skull measurements on character or mental ability. Boddy espoused a monogenetic origin for blacks and whites, and rejected the polygenism of the nineteenth century that was so often tinged with white racial superiorism. He sought to turn the tables on those who would assert inherent inferiority for African-Americans.

Boddy claimed that the dark hair and eyes of some European peoples indicated an African ancestry. "The Anglo-Saxon race sprang from the Iberian troglodytes," Boddy asseverated. For Boddy, difference in hair color or its kinky texture represented incidental results of environmental causes such as temperature. Boddy's description of the effects of heat upon the scalps of Africans was painstakingly detailed. He believed that the thickening of the skulls of those living in the tropics was due to environment, and had nothing to do with innate intellectual acumen. Boddy's claims also fit well with

---

38. Although I utilize the terminology of the discourse of the early twentieth century here it does not mean I am unaware of the long and painful history of labeling as a means of oppression. Because Boddy, during the period in which he lived, used the term "negro," I have not refrained from employing it here. My use also of "African-American" reflects awareness that this is generally the preferred term at the time this book is being written.

39. Boddy, "Brain Weight and Intellectual Development," 357.

commonplace contemporary expectations of the influence of environment as the key source of acquired characteristics.[40] He believed that African-Americans, who continued to live over several generations in the cooler climates of North America, would begin to experience a natural straightening of the hair, and would begin to appear more Caucasian.[41]

Even more complex was Boddy's explanation of the contours of the skulls of blacks, and the phenomenon of differential brain weight touted by racists as evidence for black inferiority. Those living in a "torrid zone" have thicker skulls because of their greater exposure to the sun's heat. Lest anyone were to question the veracity of his conclusions, Boddy confidently asserted: "These are cold scientific facts which are based, not upon feeling, but upon scientia scientiorum (sic) medicine and her allied sciences."[42]

Boddy cited the Egyptians, Nubians, Carthaginians, and Phoenicians as examples of advanced civilizations on the African continent. He highlighted "the more highly civilized and cultured Negroes of the pre-Christian and pre-Mosaic times." Boddy believed it likely that such peoples had skulls "no thicker than the present Anglo-Saxon's skull." If they had, Boddy took it as self-evident that "they would not have been able to erect temples and pyramids with such skill and precision." For Boddy to attribute such achievements to brain size rather than to African cultural values or education or craftsmanship indicates the influence that phrenology continued to wield in some circles at the turn of the century.

Boddy attacked many of the conclusions of the phrenologists without abandoning their project. More researches into the causes of cranial differences amongst the races had yet to be conducted, in Boddy's estimation. Racist scientists had made much of the discovery that the cranial sutures of black children converged more quickly than did those of whites. To explain this Boddy called for research into factors such as diet and metabolism among black children. Such considerations had not been adequately studied by nineteenth-century phrenologists. Therefore, all their claims "concerning the relative degree of intelligence of the Negro and the Anglo-Saxon" based on the sole criterion of cranial sutures, begged for a more just reassessment.[43] With such an argument Boddy anticipated the shift from hard hereditarian to environmental explanations of social problems that would occur with the rise of cultural anthropology in the 1930s.

40. Ibid., 358.
41. Ibid., 359.
42. Ibid., 360.
43. Ibid., 361.

For all his scientific argumentation, Boddy reflected his theological and ministerial training in that he freely employed biblical allusions throughout the essay. One particularly vivid passage struck a prophetic chord. He likened phrenological theories of Anglo-Saxon superiority to idolatry. "This suture business, then, does not constitute a racial variation," he retorted to those claiming otherwise. "They say it is, but, just as Dagon, the god of the Philistines, fell upon his face to the earth in the presence of the ark of the Lord, so that the heathen image broke into pieces," Boddy thundered, "so also, must the error of the nineteenth century fall into pieces, in the presence of the truth of the twentieth century."[44]

Boddy further delighted to point out the barbarity of ancient Gaul and Britain. Such cultures "were so low down in civilization that they would come in battle clad in skins of animals." Boddy attributed the success of Anglo-Saxons to their absorption of Roman ways, not to their innate superiority. The "arts, culture and civilization" of Rome had impelled the development of the Anglo Saxon "into the most formidable race upon the face of the earth."[45]

Boddy was optimistic that the embrace of the same cultural benefits by African Americans " in time will make the Negro race a mighty people." In this, Boddy sounded a note similar to his contemporary Booker T. Washington, who emphasized assimilation as the way to success.

Like many of his Protestant co-religionists, however, Boddy was deeply concerned about the dangers that alcohol abuse portended for the Negro race. He feared that African-Americans were in danger of following the sodden examples of Great Britain and Ireland. Boddy voiced some of the same concerns as had white health reformers such as John Harvey Kellogg concerning the "physical degeneration" of citizens of the British Isles. He referred to a 1903 study by the British Medical Association tracing this degenerative effect to "the tremendous consumption of ale."[46] Thus Boddy joined his voice with the chorus of others espousing temperance.

Boddy was a cultural assimilationist. This would have put him at odds with younger contemporaries who would a few years later promote the Harlem Renaissance and other efforts to value, treasure, and develop an unapologetic black culture. "The Anglo-Saxon civilization, with which the Negro is surrounded," he confidently affirmed, "adds powerful forces to the character of his race in proportion as he assimilates it." Far from embracing the pessimism of those white elites who predicted the extinction of blacks,

44. Ibid., 361–62.
45. Ibid., 362.
46. Ibid., 363.

Boddy believed that the embrace of Anglo-Saxon values would promote black progress. He saw the African American as improving, not degenerating. The notion of Negro degeneration and distinction was rejected firmly by Boddy as "an unchristian sentiment."[47]

Though Boddy enthusiastically promoted assimilation, he believed his position steered clear of the charge of "amalgamation." His rhetoric partook of the confusion that was endemic to the science of heredity in the early twentieth century, particular in its popularized form:

> For upwards of five hundred years the Negro has been assimilating the religion, language and other elements of the Anglo-Saxon and other races; and, at the same time, he has been gradually absorbing the Anglo-Saxon, and has made that portion he has absorbed an integral part of himself by incorporating the white "gamets" (sic) into his own race. And with it all, the Negro maintains his racial integrity and does not amalgamate.[48]

In the end, Boddy's confidence did not merely rest on racial or cultural achievement, but on theological factors. Indicative of his belief in divine approbation for African Americans were his concluding remarks: "Behind the Negro is Jehovah; He is fighting their battles for them, though unseen by mortal eye." Boddy the minister and Boddy the doctor were two identities fused in the same Boddy. The blend of scientific knowledge and biblical faith made him confident of the centrality of African-Americans in the advancement of a Christian civilization.

Some twenty-four years later, the African-American sociologist W. E. B. DuBois would debate scientific racism against the prominent eugenicist Lothrop Stoddard. By that historic moment, cultural anthropologists were laying the meticulous groundwork for attacking eugenics at its nerve center: the claim to scientific authority. DuBois was able to perceive more clearly than had Boddy the flawed elements in racist constructs of science. Still, Boddy's much less public moment of critical engagement with one aspect of racialist science, i.e., phrenology, had been an important, if transitional, phase of the African-American struggle to defeat the pseudo-intellectual underpinnings of eugenic thought pertaining to the races.[49] By contrast to Boddy's inclusive vision, Augustus Hopkins Strong would use degeneration theory to promote a striated and hierarchical approach to race.

---

47. Ibid, 364.
48. Ibid.
49. See Taylor, "W. E. B. DuBois's Challenge to Scientific Racism," 449–60.

Augustus Hopkins Strong Embraces
Degeneration Theory on Race

For Rochester Seminary's Baptist theologian Augustus Hopkins Strong (1836–1921), hereditary sin affected some groups more than others, and this opened the door to prejudicial uses of the science of heredity. Strong was not immune to the racial biases common to white elites of his era. In 1912 he published a lecture entitled "Degeneration." In this essay, Strong sought to harmonize evolution with several theological and empirical considerations.

In this essay Strong acknowledged what he called "the apparent triumph of the doctrine of evolution." For Strong, the remaining dispute was now between those who saw evolution as a blind process and those who saw it as the result of intelligence. Two forces at work in natural processes complicated this picture, but for Strong indicated the triumph of divinely-guided progress: "I claim that, while progressive evolution is the method of an immanent divine will, there is an incidental retrogressive evolution which profoundly modifies the former, and which results from a perverse human will." This retrogressive tendency Strong identified with degeneration theory.

Degeneration theory held that physiological traits of deterioration gave visible manifestation to an underlying moral deterioration in a human agent, sociological group, or even (if unchecked) the human race itself. "Civilization advances in spite of opposition," Strong announced, but cautioned that: "the stream has many a backset, temporary though the backset may be." The tension between the progressive and degenerationist ideas are discernible in Strong's claim that: "degeneration is as plain as is progress; man mars God's work, even though God overrules the evil for good."[50]

The balance of Strong's essay built a case from multiple disciplines that the struggle between the immanent divine will and human degeneration was empirically attested. In his use of history and anthropology, for example, Strong's own ethnic prejudices surfaced in theological guise in ways largely typical of his Anglo-Saxon contemporaries. "Civilization does not reproduce itself," he wrote, "It must first be kindled only by a power genuinely Christian."

Civilization served as the antithesis to degeneration in much of the literature of white elites in the Progressive Era. Civilization was very often cloaked in Christian or biblical terminology, such as "the Kingdom of God." The insidious side of civilization as an ideology was a concomitant racial

---

50. Strong, "Degeneration," 111–12.

superiorism toward those deemed outside of civilization as construed by white elites such as Strong. Rather jarring today are the words: "Modern Egyptians, Italians, and Spaniards are unquestionably degenerate races and the same is true of Australians and Hottentots, as well as of Turks." Not even Northern Europe was exempt from the theory, as Strong brought forth again an earlier generation's antipathy for the Irish, proclaiming that "the physical degeneration of portions of the population of Ireland is well known."[51] Not to leave any racial stereotype unturned, a few pages hence, Strong opined, "Both Japan and China were stagnant, if not decadent, until touched by the arts and the religion of Christian lands. Degeneration is more natural than progress, until a barbarous people comes in contact with influences from without."[52] Strong's essay on degeneration ended with his reminder to the reader that not only the Bible supported degeneration theory, but so did cultural authorities such as jurists, naturalists, historians, philosophers, travelers, theologians, anthropologists and ethnologists.

Ending in typical homiletical fashion, Strong admonished his readers, citing the Apostle Paul's theme in Romans chapter 1, that the degeneration of humanity "can be counteracted only by regeneration from above." The lingering impression of the essay as a whole, however, is that some races within the human race are self-evidently more degenerated than others.[53] Under the classical doctrine of original sin, all humans, without exception, are degenerate in some sense. Whatever the extent of that problem, it was a universal problem. In the wake of race-laden or class-laden interpretations of the science of heredity, for Strong, degeneration became less a universal problem than a special problem posed by stratification and categorization of levels of unfit persons.

Such an insular view prior to the first World War would give way to a more sympathetic analysis of ethnic differences in the period following the war. To counteract scientific racism it would require a thinker experienced in international relations and equipped with the new insights of cultural anthropology. This figure was missiologist and ecumenist Robert Speer.

Robert Speer: Missionary
Opponent of Racism

Robert Elliott Speer (1867–1947) was a highly significant figure in the history of American Presbyterianism, and of the progressive and reformist

51. Ibid., 115.

52. Ibid., 119.

53. Ibid., 128.

ethos of evangelical Protestantism. Like many pious youths of the late nineteenth century, Speer was moved to embrace evangelical Christianity by
the preaching of such prominent evangelical preachers as A.T. Pierson,
Henry Clay Trumbull, and Dwight L. Moody. As a student at Princeton
University he joined the Student Volunteer Movement and in 1887 signed
the volunteer pledge to become a missionary. Graduating as Valedictorian
from Princeton in 1889, he entered Princeton Theological Seminary, from
which he graduated in 1891. He did not pursue formal ordination, however,
serving in various administrative roles in both the Presbyterian Church,
and later in ecumenical organizations. He served as the lay secretary of the
Presbyterian Board of Missions from 1891–1937. Speer served also as the
President of the Federal Council of Churches from 1920–1924. During the
heated fundamentalist-modernist controversy, he moderated the General
Assembly of the Presbyterian Church in 1927. This term is appropriate, as
Speer consciously sought a path of moderation throughout his career, which
occasioned much criticism from those taking more firm stances on both the
left and the right.[54]

Speer's interest in race relations flowed out of both his interest in world
missions, and in his unity efforts via the Federal Council of Churches. Speer
was troubled by jingoistic treatment of both inhabitants of the Orient (e.g.,
the "yellow peril") and of African Americans (e.g., the "Negro problem") by
many Americans of Anglo-Saxon heritage. Emerging from the Great War
with a deep sense of the need to attend to domestic problems, Speer and
others in the FCC gave sustained attention to race relations. During the
war, Speer defended the rights of Japanese-Americans to acquire property
in the western states by means of his 1917 pamphlet, *How to Preserve Fellowship and Right Understanding Between Japan and the United States.* As
President of the FCC he established and participated in "racial conferences"
in January and February, 1921. Out of these meetings arose the formation
of a commission of the FCC, which by 1922 settled on the title "The Commission on the Church and Race Relations." Based on such experiences
and with the cooperation of "academicians, missionary friends and Federal
Council colleagues" Speer authored a shorter volume entitled *Of One Blood,*
and an expanded version entitled *Race and Race Relations* in the year 1924.[55]

In *Race and Race Relations: A Christian View of Human Contacts,*
Speer demonstrated the breadth of his reading in theological, social scientific, and eugenics texts on the "race problem." Speer blended realism and

54. Piper, *Robert E. Speer,* xix–xxi, 28–45, 343–57; *Dictionary of Christianity in
America,* s.v. "Speer, Robert." For a concise description of the Student Volunteer Movement, see Showalter, *End of a Crusade,* 1–7.

55. Piper, 332–40.

idealism in his approach to the issue. He acknowledged the impossibility of broaching the topic with "an absolutely and colorless mind." Nonetheless, he did claim that the Christian ought to approach the thorny issue "with what he conceives to be the mind of Christ."[56]

Speer favored the approach to the subject taken by the increasingly influential field of cultural anthropology, and the school of thought inaugurated by Franz Boas (1858–1942). Speer's enthusiasm for the methods of Boas and his followers is not surprising. Protestant missions and anthropological research have a long, albeit tense, history. For Speer the stance of such cultural anthropologists, who stressed empathetic research, and efforts to observe cultures from within as much as possible, had a special appeal. This empathetic method, with obvious affinities in the golden rule, was in Speer's estimation "to seek to behave toward men of all races as one conceives Christ would approve."[57] Perhaps cultural relativists such as Boas and other thoroughly secular students of culture would have winced at Speer's overtly Christian gloss on their thought. Nevertheless, Speer's arguments doubtless had a greater appeal for evangelicals precisely because of statements like the following:

> The thesis of this book is simply the Christian view of these questions. It holds that God made of one blood all races of men and that all races are but parts of one human race. Mankind is one great kindred of all men. That is what the word "mankind" means. In this view races are not conceived as biological

---

56. Speer, *Race and Race Relations*, 1–5.

57. Ibid., 5. Boas, a late-nineteenth-century German immigrant, earned his early fame as a student of the culture of the Eskimo inhabitants of Baffin Island in the far northwest regions of North America. After teaching at Clark University 1888–1892, he served as curator of anthropology at the Field Museum in Chicago from 1892–1899. He became a full professor of anthropology at Columbia University in 1899, where he remained until his death in 1942. Boas was a cultural relativist who viewed theories about the differential genetic endowments of races favored by eugenicists as sheer outrageous prejudice masquerading as science. He stood up again and again against discrimination against Jews, Blacks and various disfavored immigrant groups by racist elites. His books *The Mind of Primitive Man* (1911), and *Anthropology and Modern Life* (1928), as well as his editorship of the *Journal of American Folklore* (1908–1925) contributed greatly to his influence in the academic world. He published some 600 articles in his lifetime. His students represent a "Who's Who" in the annals of the social sciences, including Margaret Mead, Ruth Benedict, Alfred Kroeber, Zora Neale Hurston, and Melville Herskovits, just to mention a few. *American National Biography*, s.v. "Boas, Franz." On Boas's influence as an anti-eugenicist, see Kevles, *In the Name of Eugenics*, 134–37; and Degler, *In Search of Human Nature*, 61–104.

fixtures but simply as enlarged family groups which are subject
to moulding and transforming influences just as families are.[58]

For Speer's evangelical readers, the biblical allusion to Acts 17, and Paul's
first-century address to the intellectuals on Mars Hill was not lost. The
Biblical accounts show the Apostle seeking rapport with his audience by
affirming the essential unity of humanity. It was also a standard text cited
in opposition to the polygenism, i.e., the separate creation of various races
as distinct species, espoused by nineteenth-century scientific racists such as
Louis Agassiz and Josiah Nott.[59] The spectre of a humanity divided into a
biological hierarchy, as promoted by many prominent eugenicists, was not a
palatable one for Robert Speer.[60]

Speer clearly favored nurture over nature in that perennial debate.
Still, he reserved a place for both factors, observing that "each race is the
product not only of . . . inherited tendencies, but also of its environment
and education." Thus, if traits were not immutable and predetermined, but
were produced by some contingent processes, then such factors could alter
them in the future.[61]

Speer named specific prominent eugenicists in his criticisms of hard
hereditarianism. He rejected the hereditarian claims of Henry Fairfield Osborn (1857–1935) and Lothrop Stoddard (1883–1950).[62] Speer cited other
eugenicists, whom he regarded as more moderate on the race question, such
as Charles Davenport and Edward Grant Conklin (1863–1952). Still, for
Speer, the most promising work explaining human nature in the context of
race and race relations, was being done by "a new school of sociologists . . .

58. Speer, *Race and Race Relations*, 11.

59. On Agassiz's scientific racism, see Wallis, "Black Bodies, White Science,"
38–61. On Nott's scientific racism, see Horsman, *Josiah Nott of Mobile*, 81–103. For an
excellent survey of the history of scientific racism, particularly in its nineteenth and
twentieth-century permutations, see Hannaford, *Race: The History of an Idea in the
West*, 235–76; 325–68.

60. Other examples of Speer's citation of scripture in favor of the essential unity
of humanity included 1 Corinthians 12 and Revelation 21. See Speer, *Race and Race
Relations*, 108, 133.

61. Ibid., 14.

62. Osborn gained fame as curator (1891–1908), then president (1908–1933) of
the American Museum of Natural History in New York. Osborn was also extremely
important in the eugenics movement, most notably as founder of the American Eugenics Society and as the president of the Second International Congress of Eugenics. See
American National Biography, s.v. "Osborn, Henry Fairfield"; and Mehler, "A History
of the American Eugenics Society, 1821–1940", 409–410. Theodore Lothrop Stoddard,
a Unitarian, was a member of the Eugenics Research Association and the Galton
Society, was most infamous for his popular work *The Rising Tide of Color Against
White-World-Supremacy* (1920), cf. Mehler, 428.

which conceives human nature itself to be not a biological mechanism but a product of social relationships."[63]

Speer's convictions on the unity of humanity were only strengthened when he turned attention to new developments in biology as well. He noted that microscopic examinations of the germ plasm of both blacks and whites manifested no difference by race. A convergence of the Boasian school of sociology with developments in cytology and genetics seemed to indicate the inadequacy of purely biological descriptions of race. "In strict scientific sense there is no sure racial classification" he argued. "There is only the possibility of a broad division of human groups marked with more or less vague general characteristics of colour and inheritance and social standards and ideals."[64]

Further evidence of Speer's rejection of the main lines of hard-core eugenics may be discerned from his approving quotation of Franz Boas's *The Mind of Primitive Man*, and his citation of a scathing *New York Times* review of the xenophobic *Passing of the Great Race* by Madison Grant (1865–1937).[65]

For all his affirmations of the essential unity of humanity, Speer did not regard all races equal in achievements or in civilization. He did utilize phrases such as "inferior or backward culture" in describing groups other than his own Anglo-Saxon heritage.[66] In contrasting the cultures of Canada and Siam, it was clear that Speer held little sympathy for the Siamese, regarding their simplicity of life as "not happiness, but inertia."[67] Speer utilized the condescending metaphors of childhood when describing "others," such as "Negroes" or the Japanese.[68]

Having noted these deficits in Speer's assessment of racial others, it is still important to stress how strongly he rejected the virulent racism of nativist eugenicists such as Madison Grant and Lothrop Stoddard. Races were not by *nature* inferior and superior (even if they could be distinguished, in Speer's mind, by cultural achievements or closeness to a Christian ideal).

63. Speer, *Race and Race Relations*, 16–24.

64. Ibid., 27–30.

65. Ibid., 39–41; 106–107. Grant was a lawyer, a naturalist, and a prominent advocate of immigration restriction, see *Concise Dictionary of American Biography*, s.v. "Grant, Madison." According to historian Eric H. Vieler, Grant's *Passing of the Great Race*, translated into German, "was highly valued by the National Socialists because of its pro-Nordic racist theme." See Vieler, *Ideological Roots of German National Socialism*, 82–83; see also Kuhl, *Nazi Connection*, 73–74. See also Durst, "The Future in 1916."

66. Speer, *Race and Race Relations*, 88.

67. Ibid., 126–32.

68. Ibid., 136–37.

Speer labeled "the idea of the fixedness of race character, of the fiat of un-alterable race status" as a "radical error." Speer's cogent critique of white supremacy and xenophobic separatism warrants quoting at length:

> No race is assured of continued ascendancy. The alarmist school realises (sic) this. Indeed this is the cry of alarm it is sounding abroad. Having cherished the idea of white ascendancy it now sees that ascendancy threatened, and unconvinced of the right solution of the race problem, it is appealing for the segregation and racial withdrawal and for the eugenic race-breeding of the white peoples in the interest of the preservation of their superi-ority of race characters. Thus truth of race-growth and change is indeed a warning to all race-vanity and privilege, but it is also the hope of all races, superior or inferior. None of them is doomed to a fixed status. A true ethnological view is a confirma-tion of the promises of Christianity to the races and to the men who comprise them.[69]

For Speer there *were* superior and inferior races—but by this he meant *cultures*, including the religious component. As historically contingent phenomena, those cultures which were now "inferior" could improve, and those "superior" could (at least in theory) degenerate or slide backward. On the other hand, the view that race was an *innate* or *biological* quality had ef-fectively been rendered untenable, in Speer's view, by developments in both the natural and the social sciences.

Speer was keenly aware of the problematic nature of eugenics, and of its most recent critics in the Boasian school of cultural anthropology. Yet even Speer was not immune to the attraction of one particular form of eugenic thought, namely positive eugenics. To understand this historical oddity in his corpus, we must shift our attention some twenty years forward, and examine a short essay written by Speer in the 1940s.

According to Speer, fascination with hereditary genius was an im-portant dimension of the eugenics agenda as far back as the studies on that subject by the father of eugenics, Francis Galton. Galton's first essay on the subject, "Hereditary Talent and Character" appeared in 1864. In his 1869 book *Hereditary Genius*, along with his analysis of eminence and achievement in other professions, Galton devoted a chapter to "Divines." Here Galton had sought to give a scientific explanation for the "religious disposition." To do so he conducted detailed studies of extant biographical literature on ministers from the Reformation era to his own day. His main source was Middleton's *Biographia Evangelica* (1786). After analyzing

69. Ibid., 83.

various ill-defined descriptive facts about the lineages of several prominent Protestant ministers, Galton pronounced that: "It cannot be doubted
from these facts that religious gifts are, on the whole hereditary; but there
are curious exceptions to the rule."[70] Indeed, rigorous methodology would
have vitiated Galton's project at numerous levels. But the die was cast for
future proponents of hard heredity to trace the social ascendancy of the
children of ministers and missionaries not to environment, but to their
well-born status.

In a preface to his 1944 article on the eugenic nature of missionary
families, Robert Speer noted that the paper had originally been "prepared a
few years ago for the American Eugenics Society." Its express purpose was
"to set forth such scanty illustrative statistical data as it has been possible to
gather, with the impressions of more deliberated judgments of missionary
administrators and of educators who are in contact with missionary children in American schools and colleges."[71]

What followed was a study only slightly more rigorous than essayist
Garrison Keillor's weekly radio tribute to the inhabitants of the fictitious
Lake Wobegon, "where the women are strong, the men are good-looking,
and the children are above average." Speer's sources for his assessment of the
eugenic qualities of missionary offspring included:

1. Missionary boards of various Protestant denominations,

2. Assessments offered by various professors who had studied the missionary movement, including the famous historian of missions, Yale's
   Kenneth Scott Latourette, and

3. Statistics on the number of children in the homes of missionaries compared with those of other clergy, drawn from the 1922, and 1932–33
   editions of *Who's Who*.

It is important to observe that whereas the term "eugenics" occurred
in the title of Speer's article, it is clear that he was *not* operating on any kind
of hereditarian definition of the term. Virtually every authority Speer cited
concerning the success of missionary children stressed environmental factors: religious atmosphere, education, family size, parental example, moral
idealism, exposure to multiple languages, and the development of initiative and independence as a result of separation from parents during their
education.[72] The criticisms that Speer had leveled against hard hereditarian-

70. Galton, *Hereditary Genius*, 311–52, quoted sentence is on 326.
71. Speer, "Foreign Missions an Experiment in Eugenics," 39.
72. Ibid., 39–50.

ism twenty years earlier were far more influential in 1944 than they had
been in 1924. Palpably missing from Speer's account of the achievements
of missionary children were factors like inherited intelligence, or Nordic
germ plasm—factors earlier hard hereditarian eugenicists had seen as cru-
cial to social superiority. Speer's assessment of missionary families utterly
refrained from making concrete proposals even of the "positive eugenics"
sort typical of those active in the eugenics movement, such as explicitly en-
joining missionaries to have more children.

Conclusion

The toxic interplay of degeneration theory, scientific racism and religious
language has contributed to the lingering problems Americans experience
when faced with tensions in race relations. Major attitudinal change on the
theme of race would have to wait until the 1960s for many Americans. Over
time, social integration of the races has elevated culture and downplayed
heredity as the predominant explanation of differences in a pluralistic country. Scientific racism and eugenic thought undergirded both internal racist
social policies as well as racially-biased immigration restriction policies
toward persons of other nations. The legacy of Jim Crow laws in the South
as well as differential exposure to severe poverty between races nationwide
contributed to well-documented hardships for minorities. Supremacist and
prejudicial attitudes seem to change at an agonizingly slow pace, characterized by fits and starts and jarring setbacks. The early decades of the twentieth
century were marked by horrors such as the lynching of African Americans
on the flimsiest accusations, race riots, or jingoistic movements such as the
"yellow scare" of the first two decades. Notions of hard-wired hereditary
differences in a hierarchy of genetically inferior and superior races played a
major role in fomenting social tension and violence. The African-American
church gave voice to a new theology of equality that many institutional religions only embraced grudgingly. Other cultural forces, such as anthropology, sociology, the entertainment business, and sports did a better job of
bringing reform in race relations than did biology or religion.

   All the foregoing chapters have hinted at the need for an examination
of the theological underpinnings of discourses about human nature during the rise of biology and eugenics as culture-shaping forces in America.
The next chapter seeks a window on the ways in which biology in general, and theories of heredity in particular, impinged, both positively and
negatively, on theological reflection over the enduring question of what it
means to be human.

CHAPTER 10

Theologians, Hereditary Sin, and Eugenics

Those that we call monsters are not so to God, who sees in the immensity of His work the infinite forms that He has comprehended therein; and it is to be believed that this figure which astonishes us has relation to some other figure of the same kind unknown to man.

—Montaigne, 1588

Introduction

ORIGINAL SIN REFERS TO the Christian doctrine that Adam's sin affected not merely Adam but Adam's progeny. Christian theologians have debated the degree to which this was an inherited weakness of the soul, of the body, or of both soul and body. They have also debated whether the act of procreation itself was complicit in transmitting either inherent moral weakness or positive guilt from one generation to the next. For some original sin was primarily a constitutional weakness of humans as embodied beings. For others it was an inherited guilt in the forensic or juridical sense. Original sin was usually construed, regardless of the mode of its inheritance, as a universal problem. Theologians thus emphasized original sin in terms of the solidarity of all humans (with the exceptions of Jesus Christ and, for Catholics, the Virgin Mary) in a fallen condition. The Apostle Paul's dictum, "For all have sinned and fallen short of the Glory of God," aptly summarizes this flawed solidarity.[1]

Actual sin, however, is a distinct category, albeit rooted in original sin. Actual sin is by its nature variable from person to person. While all are sinners, not all have actually sinned to the same degree. Actual sin involves

---

1. Romans 3:23 KJV.

human choices, or acts of the will. For some theologians both original and actual sin entail guilt and moral culpability. For others guilt comes into play only upon the responsible agent's conscious choice to exercise the will in disobedience to the divine will perceived in conscience, in special revelation, or both.

Original sin, in the most popular Western form of the doctrine, was often expressed in a manner tantamount to original guilt. Original sin could be a forensic status imputed to the rest of humanity through a notion of Adam's representation of the human race. It could take the form of an inbred taint that tinctured the newly created soul at conception, in the theory known as creationism. Or it could be instantiated into the very process of the procreation of a unitary human substance in the approach known as traducianism. Thus original sin has been an enduring and contested doctrine. St. Paul's gloss on early Genesis in his epistle to the Romans serves as a central biblical datum for this discussion: "Wherefore, as by one man sin entered into the world, and death by sin; and so death passed upon all men, for that all sinned."[2] The debated question would always be: Why and how could one person's sin affect with such profound consequences all his descendents? Additional questions arose in the nineteenth century as to what role the science of heredity might play in theological discussions of multi-generational sin.

This chapter argues that by the late nineteenth century, new theories of human heredity added a layer of complexity and even distortion to discussions of sin and the soul. The classical emphasis on the universality of human original sin often gave way, particularly within the ideology of many social reformers, to eugenics, with its stress upon varying degrees of hereditary sinfulness among sub-groups within society. Such groupings betrayed a striated vision of the world in which North American Protestant elites perceived the feeble-minded, criminals, alcoholics, and immigrants as partaking of a degeneration of familial biological inheritance especially threatening to modern civilization itself. This theological shift is most obvious when elites used both degeneration theory and biblical allusions to demarcate the biologically and morally unfit from the ostensibly fitter families of the Protestant mainstream. For those theologians who embraced science enthusiastically, a desire to partake of the rising cultural authority of science was a vital component of their ethos. Eventually however some Protestant theologians began to question eugenics in important ways.

---

2. Romans 5:12, KJV. A helpful and succinct overview of the doctrine among the church fathers is Gerald Bray, "Original Sin in Patristic Thought," 37–47.

Heredity and Theology:
The Primary Sources

The analogies between eugenics and religion sit tantalizingly on the surface of the rhetoric of the eugenics primary literature. The dogma of eugenics was hereditary degeneration of problematic families, leading to the classifying of certain categories of persons as socially unfit. As an outward sign of an inward disgrace, the "stigmata of degeneration," were often represented by photographs of the cognitively disabled interpreted by negative captions. The rituals of eugenics included segregation of the sexes, marriage certificates, IQ tests, and sometimes sterilization. The sacramental substance of eugenics was the germ plasm, to be protected for the good of the whole human race, but especially for those carrying on the Anglo-Saxon heritage. Charles Davenport authored the creed of eugenics, Albert E. Wiggam gave the world the Decalogue or ten commandments of the eugenics movement.

Undergirding these ever-articulated developments was a subtle yet real theological shift from theologizing original sin as a universal problem affecting the whole human race, to an elitist focus on hereditary sin afflicting a select "unfit" subset of the population. Rather than serving a watchdog role in protecting the human rights of the "least of these," many prominent Protestant theologians either looked the other way or overtly championed the eugenics movement. Some Protestant intellectuals, such as Robert Speer and Walter A. Maier, offered criticisms of eugenics. Even these protests tended to be criticisms of fringe, extreme, or utopian modes of eugenics, or criticisms delimited to some particular ideological aspect of the movement such as racial prejudice. How this shift unfolded requires us to make a brief sojourn into the background of the doctrine of original sin as a problem of heredity in the American milieu.

Original Sin as Natural or Moral Evil

The eighteenth-century American theologian Jonathan Edwards made a fundamental distinction between natural evil and moral evil in assessing the damage of original sin to the human condition. As historian Mark Noll restates this view, humans might have natural ability, "but because people were also born sinners, they inevitably and necessarily chose self and sin over God and so displayed a 'moral inability.'"[3] A common objection offered by those opposed to the Augustinian-Calvinist emphasis on the inevitability of human sinfulness was the argument that this made God into the author

---

3. Noll, *America's God*, 272.

of sin. Edwards denied that God is the author of sin by stating that original
sin has no need of "any evil quality *infused, implanted* or *wrought* into the
nature of man" either from God or created progenitors. Rather, Edwards
held that all that was necessary to affirm the reality of original sin was to
admit that God withheld "a special and divine influence" to promote the
positive principles of the good after Adam fell. Edwards thus affirmed original sin as a corruption of nature for Adam as well as all his descendants,
whom Edwards construed "as sinning in him and falling with him."

Distinguishing the superior, divine principles of spirit from the inferior principles of the mere fleshly human nature, Edwards could account for
the fall in terms of negation. Once Adam as representative of all humanity
fell, it was as if light had been extinguished in a room: all that remained was
the darkness of an absent goodness. The inferior principles within the human person, devoid of any ennobling spiritual input, devolved into self-love
and a natural appetite running out of control. Thus it was not necessary for
God to place a positive taint or evil into the human heart for original sin to
come into existence, for the onus of sin rested on the whole human race in
their solidarity with Adam.[4]

This nuance was often lost, however, among those who would see
themselves as heirs to the Edwardsian legacy of American Calvinism. For
instance, in 1839, Philadelphia's Tenth Presbyterian Church pastor H. A.
Boardman published a sermon: *The Scripture Doctrine of Original Sin Explained and Enforced in Two Discourses*. In it he affirmed with the Presbyterian Larger Catechism:

> The sinfulness of that estate whereinto man fell, consisteth in the
> guilt of Adam's first sin, the want of that righteousness wherein
> he was created, and the corruption of his nature, whereby he is
> utterly indisposed, disabled, and made opposite unto all that is
> spiritually good, and wholly inclined to all evil, and that con-
> tinually; which is commonly called original sin, and from which
> do proceed all actual transgressions.[5]

Boardman went on to defend the doctrine of the imputation of Adam's sin
to all his progeny through his federal representation of them in the Garden

---

4. Edwards, "The Great Christian Doctrine of Original Sin Defended," 232–35.
Edwards saw Adam's relationship to his posterity in terms both of oneness in a single
covenant and an organic oneness like the relationship of the roots of an oak tree to its
branches. This bifurcation and potential for confusion between forensic and organic
metaphors will permeate the subsequent literature on original sin.

5. Boardman, *Scripture Doctrine of Original Sin*, 14.

of Eden.[6] The human race thus in Boardman's sense stood doubly damned through both a corrupt nature and a forensic condemnation.

Other Protestant Bible commentators during the late nineteenth and early twentieth century offer clues to the interplay of biblical, moral, and biological ideas of heredity held by theologians. The sermons and commentaries of London preacher Charles Haddon Spurgeon (1834–1892) of the New Park Street Church were a touchstone of transatlantic evangelical Protestant homiletics. Psalm 51 had long served as a *locus classicus* for the doctrine of original sin. In exegeting the phrase, "Behold I was shapen in iniquity," Spurgeon evoked the nexus of heredity and morality: "It is as if he said, not only have I sinned this once, but I am in my very nature a sinner." Utilizing the metaphor of illness, a standard motif in the history of discussions of original sin, Spurgeon's words took on added freight in an era of increasing medical knowledge and prestige: "The fountain of my life is polluted as well at its streams. My birth-tendencies are out of the square of equity; I naturally lean to forbidden things. Mine is a constitutional disease, rendering my very person obnoxious to thy wrath."[7] For all his subtle emphasis on biology, however, Spurgeon retained the universality central to the theological concept of original sin in its classic Augustinian expression.

Calvinism and Arminianism in American Theological Anthropology

American Protestant theology underwent a shift in emphasis from Calvinistic to Arminian assessments of theological anthropology in American theology during the course of the nineteenth century.[8] By the third generation after Jonathan Edwards, the three dominant theologians of Edwards' denomination (Congregationalism), namely Horace Bushnell, Charles Grandison Finney, and Nathanael C. Taylor had shifted away from consistent Calvinism. As Noll notes, these theologians "held that human will possessed a contrary power to act against the acquired dispositions of character and so could precipitate a change of status before God."[9] By 1900, however, the ideological pendulum had swung back toward a determinism that partook of the science as well as the theology of heredity. When the social reform

6. Ibid., 59–64.

7. Spurgeon, *Treasury of David*, 451. The biblical reference is to Psalm 51:5, KJV.

8. See Haroutunian, *Piety versus Moralism*, 220–82; and Smith, *Revivalism & Social Reform*, 88–94.

9. Noll, *America's God*, 237; cf. the discussion of Finney's critical role in this shift in Holifield, *Theology in America*, 361–68.

tradition of American theology (evident in the social gospel and temperance reforms) met hard hereditary theories, theologians struggled to find the proper balance of nature and nurture.

What has perhaps been under-appreciated is how both Calvinist and Arminian commentators came to be influenced by the nature/nurture debates in American culture, and by the insights into human nature offered by biologists, physicians, criminologists, and psychologists. The predominant tradition in the early part of the nineteenth century was New England Theology, or a modification of Jonathan Edwards's Calvinism. Among the five core assertions setting the agenda for New England Theology was the proposition: "The guilt of sin resided in the sinful choice and not in an imputation of Adam's guilt to his posterity, but sin was inevitable as a result of the fall."[10] The tension, if not contradiction, of such a proposition would emerge by the early twentieth century within both theological and scientific debates over the relative weights of nature and nurture. On the side of pessimism was the conviction that sin is inevitable, like the inevitability of certain disabilities inherited through Weismann's germ plasm or Mendel's genes. On the side of optimism stood an American frontiersman can-do attitude toward a nature routinely construed as ripe for conquest, whether the western territories rich with natural resources, or a problematic human nature corrupted by urbanization. In the following pages, several theology texts of the 1860s to the 1930s are investigated so as to illustrate the ways in which the new science of heredity made an impact on the theology of heredity, and how that impact was mediated by theological categories.

Joseph Van Dyke and the Problems of Heredity

In the struggle to accommodate the findings of science into the ministerial teaching vocation, historians have noted a wide spectrum of views. Some, such as Princeton Seminary theologian Charles Hodge, published his *What is Darwinism?* in 1874, and rejected Darwinism as atheism. Others embraced the Darwinian paradigm enthusiastically, simply asserting that evolution was God's chosen method of creating variety on earth. Most thinkers sought a place to stand in the gray middle expanse separating these extremes.

Taking his place in the murky middle, Joseph S. Van Dyke (1832–1915) may be described as a half-hearted endorser of the theory of evolution. A graduate of Princeton Theological Seminary, a Presbyterian pastor, and an occasional tutor at Princeton, Van Dyke published the second

10 Holifield, *Theology in America*, 343.

edition of his *Theism and Evolution* in 1886. Though a student of Charles Hodge, Van Dyke departed from his mentor's strongly anti-evolutionary line, yet maintained a pervasive skepticism toward many claims of the evolutionists in his day.[11]

Most relevant to our subject matter was Van Dyke's chapter: "Darwin's Primordial Germs." The prose in this essay reflected the wide ranging and vague terminology endemic to the science of heredity in the 1880s. Van Dyke used this uncertainty as an apologetic tool in order to raise many questions and objections relating to the notions of heredity then current. One of the indistinct terms popular in science was "protoplasm." He called into question the view that "Man's parentage ought to be traceable backwards through the vegetable kingdom, for plants are protoplasm, indistinguishable, it is said, from human protoplasm."[12] Trying to protect human uniqueness, Van Dyke questioned the standard accounts of an evolutionary pathway preceding human emergence.

Seeking the specific pathway from lower life-forms to humanity, Van Dyke evinced frustration with the lack of definitive answers in science. Van Dyke referred to the field of embryology, noting that "the human embryo, it is affirmed, differs in no respect, in its earlier stages, from the embryos of other animals." He would go on to devote an entire chapter to arguing against Haeckel's reliance on spontaneous generation of a simple primordial life form (the "moneron") as the basis of evolution. Van Dyke expressed curiosity about the cause of the branching or differentiation, and noted that either multiple intelligent creations were necessary, or else an embrace of an oft-discredited spontaneous generation. What he could not conceive of was "simple germs, ninety per cent water, and differing in no respect from ordinary germs," having within them "concealed powers adequate to the task of developing" into the widely divergent species observed by science.[13] Here Van Dyke betrayed a lack of sufficient awareness of the nascent field of cytology, where the profound complexity in the cell, as well as differentiation between somatic and germ cells, was beginning by this time to be appreciated by scientists.

What bothered Van Dyke, and continues to roil over a century later in the debates over intelligent design, was the upward movements of complexity attributed to unintentional (or ateleological) forces such as natural selection and sexual selection. Van Dyke complained about the demands of differentiation, namely: "how can a single substance proceed to become

11. Moore, *The Post-Darwinian Controversies*, 241–45.

12. Van Dyke, *Theism and Evolution*, 127; see 145–60.

13. Ibid., 131.

complex?" Assuming, with Darwin, that "primitive germs" were relatively simple, Van Dyke was mystified by the complexity of the process of variation.

Not averse to using sarcasm toward atheistic interpretations of evolution, Van Dyke made statements such as: "Having, for some inexplicable reason, determined to develop into animals, these germs must next decide which way to evolve, upwards or downwards."[14] In philosophical terms, this was the problem of teleology, especially the role of final cause, or purpose, in natural history. It was not uncommon for proponents and critics alike to urge a dysteleological, or undirected, interpretation of the scientific data of change in the natural world. Regression was a decidedly dysteleological concept in so optimistic an era as the late nineteenth century. Van Dyke embarked on a discussion attempting to account for the oft-noted phenomenon of regression. For example, he observed that "as the complex frequently degenerates into the more simple" and that "degeneration was possible, or at least improvement was not certain" in natural history. Showing his skepticism toward the popular progressivist vision of evolution taken by some Protestant commentators, Van Dyke averred that "retrogression is nearly as frequent as progression." This he perceived as a problem for the earliest life forms, commenting that "it is strange, then, that in the initial period, evolution, in every instance, was towards higher forms."[15]

Honing in on the problem of reversion, Van Dyke invoked a trope that, when later extrapolated to human society, would strike fear into white Protestant elites. "The tendency to revert to ancestral forms is as powerful as the tendency to preserve increments of improvement; and the tendency may lie latent in organisms for thousands of generations." Here he quoted Darwin's *Animals and Plants under Domestication* to the effect that: "this principle of reversion is the most wonderful of all the attributes of inheritance."[16] In a Victorian context, "wonderful" connoted an amazing, puzzling, or perplexing idea difficult to accept, and did not carry the positive overtone it does today. Van Dyke sought to interact with scientists in his time, even if it meant expressing skepticism at every turn. He believed he could cite reversion as a way of questioning the then-popular progressive interpretation of evolution. While Van Dyke adopted a strongly critical and lukewarm reception of evolution, his contemporary George Frederick Wright had a more variable interaction with Darwin's theory over a long career in both science and theology.

14. Ibid., 134.
15. Ibid., 135.
16. Ibid., 137.

George Frederick Wright, Calvinism and Heredity

George Frederick Wright (1838–1921) as a youth was heavily influenced by the evangelical revivalism and perfectionism of famous evangelist and social reformer Charles Grandison Finney. As a collegian, Wright studied at Oberlin College in Ohio, an evangelical college Finney had served as co-founder and president.[17] After completion of theological courses, Wright served as a Congregationalist minister to congregations in Bakersfield, Vermont and Andover, Massachusetts.

During this period he engaged in investigating local geological and glacial phenomena. He also became a correspondent and confidant of the famed Harvard botanist and prolific theistic evolutionist Asa Gray.[18] Like Gray, Wright, for most of his life, believed possible a harmonization of the Bible with Darwin's transmutation hypothesis. This optimism would change around the turn of the century.[19] In 1881, Wright accepted a professorship in New Testament Language and Literature at Oberlin College. In late 1883, Wright took over the editorship of the evangelical journal *Bibliotheca Sacra*.[20] Wright's tenure as editor of *Bibliotheca Sacra* spanned 38 years—the longest editorship in its history. This position gave him a platform for espousing his views on the interaction between science and theology.[21] Wright remained editor of *Bibliotheca Sacra*, and a professor at Oberlin, until his death in 1921. This organ serves as a source for many of Wright's publications on theology and science.

Because Wright worked in the two realms of science and theology, he had a unique perspective on how a theologian ought to understand and engage with science. Early in his career, in his 1878 essay "The Proper Attitude of Religious Teachers Towards Scientific Experts," he urged that theologians

17. Numbers, "George Frederick Wright," 624; see Morison, "George Frederick Wright," 1–98.

18. See Wright, "The Debt of the Church to Asa Gray," 523–30.

19. Numbers, "George Frederick Wright," 625–26; Morison, "George Frederick Wright," 99–149.

20. Lippy, *Religious Periodicals of the United States*, 62. *Bibliotheca Sacra* is the American religious journal with the longest life, having been published continuously since 1843. *Bibliotheca Sacra* became the prime instrument in the late nineteenth century of the moderated Calvinism of the New England theology. From 1852 to 1884, Edwards Amasa Park served as the journal's editor-in-chief, and it was Park who orchestrated both the organ's move to Oberlin, and Wright's assumption of the position of editor-in-chief. Bennetch, "The Biography of Bibliotheca Sacra," 8–12.

21. For an excellent overview of the changing relationship of science and scripture within Protestant intellectual circles, see Roberts, *Darwinism and the Divine in America*, 91–242.

"should be extremely careful to abstain from acting as umpires between scientific disputants." He used the rhetoric of "fences" and "boundaries" between the disciplines, but acknowledged that these are not always clear, and that border conflicts do occur. His own incursions into science gave him the right, in his view, of passing "intelligent judgment on the credibility of witnesses who report scientific observations, and upon the bearing of their established facts upon a theory of causation."[22] This was an early example of his understanding that science and theology are distinct, but there is no impermeable boundary between them. Wright's apologetic strategy hinged upon blending the two disciplines.

In 1882 Wright devoted an entire chapter to the topic of analogies between Darwinism and Calvinism in his *Studies in Science and Religion*. In this essay he used Calvinism and evangelicalism as synonyms, indicating an effort to speak for a broad, multi-denominational tradition. The key affinity Wright perceived between Darwin and Calvin was the "doctrine of the continuity of nature." For Wright, the theological version of this concept came from the 1646 Westminster Confession's assertion that "God hath foreordained whatsoever comes to pass." To make common cause with science was important to a churchman/scientist such as Wright. "The modern man of science," he confidently claimed, "in extending his conception of the reign of law, is but illustrating the fundamental principle of Calvinism."[23]

Wright pointed out that Darwin's theory of natural selection contained both advancement and degradation among species. This Wright applied to human beings, especially the concept of degradation through the doctrine of the fall.[24] Wright went on to analogize "the Calvinistic doctrine of the spread of sin from Adam to his descendants" with "the Darwinian principle of heredity."[25] Adopting Calvin's metaphor of corrupt branches (humanity) emerging from a corrupt root (Adam), Wright quoted Calvin to the effect that individual humans contract vice "by a hereditary law." This metaphor led Wright to an extended discussion of traducianism, the notion that human souls are all really one soul branching out over human history from Adam. Here Wright was coy, refusing to come down on the side of direct creation of the soul or traducianism. He dismissed the debate by concluding: "The naturalist, as such, is not compelled to settle questions in theology

---

22. Wright, "The Proper Attitude of Religious Teachers," 781. Cited in Numbers, "George Frederick Wright," 627.

23. Wright, *Studies*, 220.

24. Ibid., 221.

25. Ibid., 224.

and metaphysics."[26] A reading of Wright's broader body of work shows that he used this evasion technique often when a sticking point between theological and scientific discourses seemed to arise.

Wright's center of gravity shifted away from Darwinism by the early twentieth century. In *The Fundamentals*, the chief articulation of conservative Protestant beliefs of the era, Wright had not noticed some major advances in the science of heredity. In his 1913 contribution to the famed essays designed to fight modernism, "The Passing of Evolution," Wright claimed:

> This mysterious capacity for variation lies at the basis of (Darwin's) theory. If anything is to be evolved in an orderly manner from the resident forces of primordial matter it must first have been involved through the creative act of the divine Being. But no one knows what causes variation in plants or animals. Like the wind it comes, but we do not know whence it cometh or whither it goeth. Breeders and gardeners do not attempt to produce varieties directly. They simply observe the variations that occur, and select for propagation those which will best serve their purposes. They are well aware that variations which they perpetuate are not only mysterious in their origin, but superficial in their character.[27]

The historical reality was that in 1913 breeders and gardeners had been attempting to produce varieties directly, and making numerous breakthroughs for several years. While the variations may have been superficial in morphological terms, the grasp of genetic mechanisms of change among scientists was deeper than Wright appears to have appreciated. The "mysteries" of heredity were beginning to receive detailed scientific description by geneticists by the year 1913. Theologians younger than Wright who were more enthusiastic about the blending of science and theology were emerging on the scene. Influential in the field of Protestant theology, their names were Newman Smyth and Shailer Mathews.

Smyth and Mathews Wed Science and Religion

In 1913, Andover-Newton Seminary theologian and ethicist Newman Smyth (1843–1925) delivered a series of lectures under the auspices of the

26. Ibid., 226–30; quote from 230.

27. Wright, "The Passing of Evolution," 16. The wind metaphor was originally a reference to the mysterious workings of the Holy Spirit in the New Testament Gospel of John, see John 3:8.

Taylor Foundation at the Yale University School of Religion. Smyth's address to highly-educated seminarians of the era offers a vital window on science-religion relations in the Progressive Era. That a theologian should surrender significant ground and authority to scientific method may surprise us, given that the warfare model still holds powerful sway in the media and in social consciousness generally. But the rethinking of theological verities in the light of science has seen multiple generations of iterations over time.

In *Constructive Natural Theology*, the book that derived from Smyth's Yale lectures, Smyth opened with a poem by that late-nineteenth-century scion of elite northeastern poets, James Russell Lowell:

> Science was Faith once; Faith were Science now,
>
> Would she but lay her bow and arrows by
>
> And arm her with the weapons of the time.
>
> Nothing that keeps thought out is safe from thought.
>
> For there's no virgin-fort but self-respect,
>
> And Truth defensive hath lost hold on God.[28]

For Smyth, as for Lowell, the weapons of traditional apologetics or scripture must be laid aside, and a peace treaty brokered with modern science. As Smyth's argument proceeded, he called upon theologians to reconstruct their theology in light of scientific disciplines such as physics and biology, as well as history, biblical criticism, psychology and sociology—all cast within a modern scientific mold. Lamenting the dearth of natural theology in schools of divinity, Smyth called for "a theology of nature, constructed in accordance with known principles of evolution." Anything less would render religion immature, "a child's fancy thrown lightly out upon the mystery of the world." Such a new natural theology had to be forged "from the ascertained data of natural science."[29]

Smyth called upon his Yale divinity audience to approach reality with the virtue of humility, in rather maudlin terms, stating: "I am thinking of the genuine man of science, of the man who will not deny his own intellectual devotion to truth by failing to keep a heart as reverent and as humble as that of the simplest believer who looks up with worshipful eyes to the Madonna and the Holy Child . . . ."[30]

At the climax of Newman Smyth's argument the metaphor of the wedding of science and religion seemed to pass beyond mere metaphor, as he

28. Smyth, *Constructive Natural Theology*, vii.

29. Ibid., 3.

30. Ibid., 108.

rhapsodized: "Scientific devotion, kept unbroken until death, is the troth of a man's being to God's truth . . . ."[31] The marriage he envisioned was more of an arranged than a companionate one, however. For the subtext of Smyth's argument implied that theology should graciously and humbly submit to science within a union of unequals. This process would involve the blending of Protestant religion with the new science of eugenics through the influence of clergy and their educators.

In the Progressive Era, the American Eugenics Society set forth a call for ministers to enter their Eugenic Sermon Contest. Now available online, select sermons from this series shock the modern reader with the clergy's enthusiastic embrace of the eugenics movement in the USA. Such strange bedfellows as the Protestant, mostly mainline denominational clergy, and the well-funded eugenics movement bespeak the complexities of the history of eugenics.[32]

An embrace of eugenics was not merely evident among clergy, but amongst their theological educators as well. Indeed it is arguable that the warm embrace of many mainline denominational clergy of the concept of eugenics has roots in their ministerial education. Dean of the Chicago Divinity School, Shailer Mathews (1863–1941), wrote regarding many of his former students preaching in pulpits and teaching in seminaries by the late 1930s. He conjectured that: "It is probably correct to say that generally they represent a realistic view of religion and I like to think that our emphasis upon a scientific approach to contemporary individual and social needs has helped forward a morally vital rather than a merely doctrinal conception of Christianity."[33]

Mathews edited the 1924 volume entitled *Contributions of Science to Religion*, in which the chapter promoting eugenics was authored by none other than Charles B. Davenport, President of the American Eugenics Society. Mathews himself wrote of the corruptions of human nature as: "those passional elements which humanity shares with other animals." Mathews suggested that such elements were what "Augustine had in mind when he spoke about original sin." For Mathews, the great ancient moralists such as Plato, Paul and Augustine lacked science, and this led them "to very imperfect and sometimes grotesque explanations of the facts." Elevating science to something akin to a theological truth, Mathews added: "But it is not hard to

31. Ibid., 118.

32. The key repository for the American Eugenics Society Eugenic Sermon Contest sermons is the American Philosophical Society in Philadelphia, Pennsylvania.

33. The University of Chicago Divinity School, founded by the largesse of John D. Rockefeller, represented this attitude toward the authority of science. Mathews served as dean of the school for many years. Mathews, *New Faith for Old*, 268–69.

see how sympathetic Augustine might have been with our modern knowledge of evolution and eugenics."[34]

The linkage of such a mentality to the rise of positivism in the sciences, most notably in the social sciences, is ably explored elsewhere.[35] The embrace of eugenics by many prominent theologians and clergy serves as further evidence of such trends. When it came to the reception of Darwinian evolution among religious intellectuals in America, as historian Jon Roberts has documented, the response was sharply divided. The majority became accommodationists who found ways to adjust their theological convictions so as to incorporate insights from science into their belief-systems. But a vocal group demurred, namely a "sizable minority of religious thinkers," i.e., religious leaders who came to "view the theory with a jaundiced eye." They were concerned that such an embrace of evolution would reduce the sense of divine providence in the daily lives of believers, and that such a concession would undermine biblical authority.[36] This struggle would crystallize in the very public 1925 Scopes Trial over the teaching of evolution in the public schools. One figure who tended to see degeneration as a significant threat to progressive religious reform was Luther Townsend.

Luther Tracy Townsend's Pessimism

Degeneration theory's impact on theology is illustrated in the writings of Luther Tracy Townsend (1838–1922). Townsend was a Methodist clergyman who wrote anti-evolutionary evangelical apologetics. Townsend was a professor of church history, classical languages, and theology at (what is now) Boston University School of Theology from 1868 to 1893.[37] Based on a December 1904 lecture delivered before the American Bible League, Townsend published in 1905 and again in 1921 his book *The Collapse of Evolution*. One of the points of contention he raised was his belief that human degeneracy falsified any progressive interpretation of evolution. In a chapter entitled "Ancient Civilization," Townsend decried the downward spiral of civilization he saw in evidence in his own time. "Throughout the world there is evidence of a brute violence such as was unknown in the ancient empires," Townsend railed. He cited the warden of New York's Sing

34. Mathews, *Contributions of Science to Religion*, 412–13.

35. See especially Cashdollar, *Transformation of Theology*, 142–81.

36. Roberts, "The Struggle over Evolution," *Encyclopedia of American Cultural and Intellectual History*, 1:592–93.

37 *Concise Dictionary of American Biography*, s.v. "Townsend, Luther Tracy"; see also Moore, *Post-Darwinian Controversies*, 198–99.

Sing prison, Lewis E. Lawes, to claim that the modern prison represented "a phase of degeneracy and perversity with which the world hitherto has not been familiar."[38] Townsend saw "in this trend of ethical degeneracy" (specifically criminality) another social problem, namely: "the increase of mental and physical unfitness." He tied a sharp increase in suicides in China, England, France, Germany and the USA to "the increase of criminality and mental degeneracy."[39]

Townsend blamed evolutionary theory and the writings of Friederich Nietzsche, who had denounced Christianity, "as being a system calculated to make more degenerates out of men," among other putative faults. Townsend warned his readers that if evolution gained more influence the result would include increased warfare, murder, and that the world would become: " . . . an asylum for an idiotic race and a mad house with padded cells without anyone to lock or unlock its doors." With sermonic vividness, Townsend played upon the same fears that eugenicists (either religious or not) tended to stoke. "The last man, a degenerate, will curse God, dying with the curse on his pallid lips," Townsend thundered. Calling on modern citizens to repent of their various sins, Townsend presaged the end of the world: "Such is the world's prospective doom unless there is a supernatural interposition."[40]

While such lurid language was not unusual in some areas of theology/science polemics, the point to be made is that tropes of degeneracy were highly malleable. For some, biology was the culprit for human degeneration; for others environmental sin or vice was primarily responsible; for still others erroneous ideologies led to degeneration. For many, the preferred strategy was to blend a variety of causes in an explanatory matrix so as to warn, Jeremiah-like, of an impending doom for civilized humanity. The ability to strike a balance between utopian progressivism and dystopian degenerationism was a skill rather rare, either among theologians or scientists. Indeed, William James had written and lectured extensively about this bifurcation in religious thought in his *Varieties of Religious Experience* at the turn of the century. The blending of scientific and theological sources of knowledge took an even more public form with the advent of mass media such as radio. One theologian arose in the 1920s and 30s who addressed the claims of science as a true public theologian. Our discussion now turns to his writings because he wrote and preached often about eugenics. His name was Walter A. Maier.

---

38. Townsend, *Collapse of Evolution*, 61.

39. Ibid., 63.

40. Ibid., 64.

Walter A. Maier and the
Critique of Eugenics

Walter A. Maier (1893–1950) was a leading intellectual figure addressing evangelical young adults of marriageable age during the interwar period. Maier was a pastor, a Harvard-trained Old Testament professor, and pioneer in radio evangelism for the Lutheran Church Missouri Synod (LCMS), the church in which he was raised. His initial method of reaching a wide audience was through his assumption of the executive secretary position in 1920 of the Walther League, the leading youth organization of the LCMS. Maier edited for some 25 years the League's official organ *The Walther League Messenger*, which tackled numerous theological, social, and ethical issues germane to youth and young adults. During his tenure as editor, its circulation went from 7,000 to 80,000 subscribers.[41]

Maier earned his PhD in Semitics at Harvard University in 1929.[42] His teaching career as a professor of Old Testament at Concordia Seminary, an institution of the confessional and theologically conservative LCMS had begun in the early 1920s. By the middle of the twenties Maier began to recognize the potential of radio for disseminating Lutheran teaching to a wide audience. He was instrumental in establishing a radio station, KFUO, in the attic of the seminary. Its 500-watt transmitter sent forth its first broadcast on December 14, 1924—early in the history of radio.[43] The Lutheran Laymen's League in 1930 asked Maier to speak on their nascent radio forum "The Lutheran Hour." In 1935 Maier became the highly popular program's established voice, a position he held until his death in 1950[44] Maier's writings were often targeted to a broad audience, giving him a platform to address social policies in a more fulsome manner than the systematic theologians surveyed above.

---

41 *American National Biography*, s.v. "Maier, Walter Arthur." Historian of American Lutheranism E. Clifford Nelson notes that the Walther League was founded in 1893 with the following aims: "the inculcation of loyalty to the church, the promotion of local societies, their preservation from affiliation with heterodox societies, and the promotion of sociability within and among societies." The Walther league struggled for most of its first two decades, but, according to Nelson, "after the upturn in its fortunes which dated from 1910 it developed into a thriving and effective organization." Doubtless much of this success could be attributed to Maier's efforts. Nelson, *Lutherans in North America*, 302–303. For brief discussions of C.F.W. Walther and the origins of the Missouri Synod, see Gritsch, *History of Lutheranism*, 194–99; and Wentz, *Basic History of Lutheranism in America*, 202–20.

42. Maier, *A Man Spoke, The World Listened*, 55; 85–92.

43. Ibid., 72.

44. Ibid.

James D. Bratt has described Protestant immigrant communities in-cluding LCMS Lutheranism as "continental in origin, confessionalist in the-ology, Restorationist in program, and tribal in culture and social structure." This generalization does not adequately describe a Walter Maier who was eager to break his denomination out of social isolation.[45] Maier's qualified and hedged criticisms of eugenics are best seen as a product of his defense of the cultural authority of Protestant ministers based in biblical concerns, in anti-Catholicism, in his own inner struggle between elitist and populist im-pulses, and in a selective and ambivalent appropriation of secular experts in the biological and social sciences for his conservative Lutheran apologetics.

In his radio messages, in his articles in the youth-oriented magazine the *Walther League Messenger*, and in the eventual marriage manual based thereon, Maier set forth his views on a wide range of issues connected to marriage. Maier's articles, radio messages, and books formed a tacit apolo-getic for conservative Biblicism, orthodox Christology, and Protestant ministerial authority. He critiqued the marriage advice of various profes-sionals not merely for factual accuracy but for flaws he perceived in the philosophical or theological assumptions driving that advice. Such criticism then opened the door for Maier's preferable pastoral directives.

Maier's condemnation of human sinfulness, understood as innate via original sin, and ever proliferating as a result of actual sins, was a hall-mark of his conservative Lutheranism. The most damning sin of all was human pride, which Maier colorfully skewered in a wry essay criticizing a sensational laundry list of nuptial innovations in the roaring twenties. In his 1927 article "Here Comes the Pride!" Maier quipped that: "June brides and grooms must realize that the sacred ceremony of marriage must not degenerate into an ostentatious display of clothing, a depressing orgy of extravagance, or a sensational bid for notoriety."[46]

In the thirties Maier's published radio messages bear the marks of an increasing sense of urgency about the state of the home. Concordia Pub-lishing Company regularly published Maier's sermons in annual volumes through most years of the thirties and beyond. Perusal of these volumes shows that messages on marriage, family, and home life were featured at least four times per broadcast season. In a 1936 message entitled "Build the Home with God," Maier decried the moral decline of the American home, and criticized the burgeoning social sciences as ineffectual to stem it.

45. Bratt, "Protestant Immigrants and the Protestant Mainstream," 111. By contrast, see Walter A. Maier's efforts to move his denomination more visibly into the public square in his article "Away with our 'Splendid Isolation'!," 112.

46. W. Maier, "Here Comes the Pride!,"536; see also W. Maier, "Worse than Plung-ing Parachutes," 266–67, 319.

"The physicians, psychologists, psychoanalysts, sociologists and biologists who have been drawn into hurried consultation have effected no definite improvement," Maier asserted. He concluded that: "Their efforts, as beneficial as many are in a secondary way, cannot meet this emergency." Maier's solution was to quote Psalm 127: "Except the Lord build the house, they labor in vain that build it."[47] Similar expressions of a kind of ministerial and populist skepticism toward cultural experts are sprinkled throughout Maier's corpus.[48]

Maier wrote frequently on the theme of eugenics in the *Walther League Messenger*—usually in pastoral admonitions regarding marriage aimed at a young adult audience. His critical stance toward secular marriage advice manuals and his media savvy style led him to address the eugenics movement in American culture. In his 1934 essay "Must Marriage Go Eugenic?" we find encapsulated the arguments Maier would later develop in book form. Maier noted with special alarm the "far-reaching and radical platform of marriage reform" of the early Hitler regime. Maier lambasted the eugenics movement in its American iteration as well. He recognized eugenics in two forms: positive and negative. To negative eugenics, Maier was cautiously open. Citing a passage of the Jewish law demanding quarantine for lepers, Maier claimed that the church "has never felt itself called upon to protest against state laws which prohibit the marriage of epileptics, feeble-minded or insane persons."

Maier was far more critical of positive than negative eugenics. His grounds for concern lay in a variety of philosophical and quasi-religious premises that fed what he labeled the "cult" of eugenics. For example, Maier recoiled from applying to humans the same principle of selective breeding farmers typically applied to their livestock. He saw in this a mockery of the divine role in child-bearing, citing the biblical refrain, "Children are an inheritance of the Lord." He regarded selective breeding as an untenable form of biological determinism that unduly excluded divine influence.

At least five other objections rounded out Maier's contemptuous analysis of the "cult" of positive eugenics. First, he disavowed the program as prizing physical characteristics over spiritual virtues. Secondly, Maier believed eugenics was built on a premature and uncertain foundation, noting that "the study of genetics, the science of heredity, is still in its formative stages." Thirdly, he believed that eugenics had the potential to be "a cruel instrument of tyranny." In the fourth instance, Maier refused to privilege heredity over

47. W. Maier, *Christ for the Nation!*, 49.

48. See W. Maier, *Fourth Lutheran Hour*, 82, 209; W. Maier, *The Cross from Coast to Coast*, 161; W. Maier, *Radio for Christ*, 61–62, 228, 236.

environment as did many eugenicists. And finally, Maier utilized withering sarcasm to lampoon the "absurd extremes" the eugenic movement had taken in various transatlantic manifestations. Citing among others Richard Chapman's eugenic novel *A Vision of the Future*, Maier warned of eugenics as a marriage-destroying force.[49]

Maier's arguments over eugenics involved a paradoxical appeal both to elitist and to populist sentiments. He could also sound forth both pro-catholic and anti-catholic sentiments. In a theological sense, Maier was still representative of an orthodox, confessional Lutheranism that rarely let pass an opportunity to criticize Catholicism. In a sociological sense, Maier was representative of an immigrant community that only a decade before been persecuted during the nationalist fervor of the First World War, and had made Herculean strides to demonstrate a staunch American identity.[50] Along with many in the Protestant mainstream, Maier saw in Catholicism a perceived threat to American democracy, a sentiment widely evinced by the harsh pan-Protestant opposition to Catholic Al Smith during the presidential campaign of 1928.[51]

Maier had multiple mutually reinforcing motivations to focus his criticism on Catholic priests, and he saw celibacy as an opportune opening and a tender point at which eugenicists similarly often leveled criticism against the Roman Catholic Church. In a section lauding the blessings of a married clergy, Maier pointed to eugenic studies that identified the offspring of ministers with the upper echelon of elite culture. He quoted approvingly from the 1930 address of Dr. Clarence G. Campbell, president of the Eugenics Research Association: "We know from reliable, factual data that the best quality of leaders rises, and rises in the greatest frequency, from the progeny of the clergy." Maier proceeded to document many prominent Americans drawn from ministerial stock, including eight signers of the Declaration of Independence, the Wright Brothers, William James, Ralph Waldo Emerson, Henry Clay, Samuel F. B. Morse, and Oliver Wendell Holmes,[52] the latter the author of the infamous *Buck v. Bell* opinion asserting that "three generations of imbeciles are enough"[53] Such lines of argument are also present in many of the Eugenic Sermon Contest sermons in the American Eugenics Society archives in Philadelphia.

---

49. W. Maier, "Must Marriage Go Eugenic?," 328–29.

50. Bratt, "Protestant Immigrants and the Protestant Mainstream,"123–24.

51. See Fisher, *Catholics in America*, 103–107.

52. W. Maier, "Here Comes the Pride!," 30–31.

53. Kevles, *In the Name of Eugenics*, 110–12.

Maier contended against the Catholic position that celibacy is a state superior to matrimony. He did so citing biblical grounds, arguing that both Old and New Testaments promoted marriage, that Paul's recommendation of celibacy was of tentative and temporal application, and that Paul also declared that "A bishop . . . must be blameless, the husband of one wife."[54]

Drawing upon the very sorts of experts he often critiqued, Maier gleaned support from prominent eugenicists (and a long lineage of Protestant apologists) to assert that clerical celibacy was not only unbiblical, but also subversive of society:

> Suppose for a moment that the ideals of celibacy were carried through to their logical extremes, that a program of matrimonial anarchism could be enforced and that in revolt against Bible standards Christian marriage were utterly obliterated. The resultant picture would be hideous beyond imagination.[55]

Maier's reliance on experts in the rising academic specializations of the early twentieth century was not, however, unqualified. In his chapter "Marriage from God—Not from the Gorilla," Maier decried those biologists who stressed a "brute beginning" for the institution of marriage. He perceived such a tenet as morally disastrous in its implications, for, if true, "then we must ultimately be ruled by animal regulations and governed by brute impulses." With blistering scorn, Maier declaimed:

> If the love that binds husband and wife together and builds the Christian home as the haven of earth's highest joys is but the refinement of animal rut, an evolution from the biology of the beast, then away with marital law and order! Down with decency and purity and constancy! Let conscience perish and conventions crash! Abandon restraint! Give us a perpetual carnival of promiscuity, lifelong Saturnalia of sin![56]

On the other hand, Maier could cite biological findings approvingly when they undergirded an argument he sought to make. For instance, in a section criticizing modern birth control movements, Maier cited one of the most prominent biologists in the eugenics movement, Henry Fairfield Osborn. By 1939 Osborn was the late director of the American Museum of Natural History in New York City, who had decried the dysgenic side effects of the birth control movement. While eugenicists generally favored birth control for the "lower" or "inferior" classes, they realized that the educated

54. W. Maier, "Here Comes the Pride!," 27–28.

55. Ibid., 31.

56 W. Maier, *For Better, Not for Worse,* 73.

elite classes were far more likely to practice birth control, and thus reduce the numbers of their progeny in the populace.[57] Maier quoted approvingly the dire warnings of "race suicide" by eugenicists Ellsworth Huntington of Yale, Leon F. Whitney, executive secretary of the American Eugenics society, and Edwin Grant Conklin of Princeton. He occasionally voiced similar warnings in his radio sermons.[58]

Elsewhere in his marriage manual Maier tempered this apparent elitist line of argument with a note of populism, thus manifesting the tension between populist and elitist impulses in his own cognitive framework. He criticized as "misguided" those proposals that "would ultimately restrict parenthood to the wealthy and prohibit those unable to support children from reproducing." Labeling this as a "spirit of discrimination," "absurd," "unAmerican," and "unchristian," Maier offered a litany of luminaries in American history who came from impoverished families yet achieved great success as national leaders.[59]

Maier's skepticism toward scientific experts led him to regard statements of certain sociologists with misgivings. In his chapter "The Sociological Nightmare" Maier offered warnings about sociology as a stalking horse for such evils as atheism, anarchism and free love. These he saw as "radical sociology" and as an attack on traditional marriage.

Maier demonstrated such resistance in his evaluation of sociological proposals of birth control as a panacea for most social ills, including the much-feared proliferation of the mentally disabled, the alleged "menace of the feebleminded" popularized by H.H. Goddard, Charles Davenport, and a host of other eugenicists.[60] Maier's argument regarding the so-called "feebleminded" was markedly pragmatic rather than explicitly theological, and his rhetoric about genetically suspect persons still echoed the strains of degeneration theory:

> Every intelligent observer should concede that degenerate, mentally defective men and women, of whom medical science can definitely state that their offspring will be similarly deficient, should not be permitted to propagate. However, the application of this axiomatic principle may entail several almost inextricable difficulties. Where shall the line of elimination be drawn? Hereditary and eugenic research has shown the futility of absolute

57. See Reed, *Birth Control Movement and American Society*, 134–36, 208.

58. Maier, *For Better Not for Worse*, 380, 402–403; cf. W. Maier, *Lutheran Hour*, 278, 284, 294; W. Maier, *Christ for the Nation*, 51.

59. W. Maier, *For Better Not for Worse*, 388.

60. Kevles, *In the Name of Eugenics*, 77–80.

rules. . . . Eugenics knows many deformities not communicable
by heredity. Even if these border-line parents could be definitely
warned that their progeny would be defective, birth control
would still be dangerous, because its methods often fail.[61]

The specter of rampantly reproducing "degenerates" clearly strained Maier's
social sensibility. That "they should not be permitted to propagate" made
sense to him, but only if such a program of restriction could *succeed*. This
was largely a pragmatic argument, and since such persons could not effec-
tively employ birth control, efforts to coerce them into its practice Maier
simply deemed to be futile.

For Harvard-educated Walter Maier any temptation to elitism was
tempered by a respect for the common person. As a pastor who interacted
with all levels of society, who traveled and spoke widely, and who carried on
a popular radio ministry, a purely elitist mentality was virtually impossible.
His radio ministry, as he noted often in his radio sermons, was backed by
the Lutheran *Laymen's* League. But the desire to be counted an expert, a
cultural intellectual, and to reinforce cultural authority with scriptural authority and Protestant hegemony, created some rhetorical flirtations with
negative eugenics. On the substance Maier was a skeptic toward eugenic
claims. Maier was critical of negative eugenics as public policy, based on
doubts about the scientific basis or social efficacy of sterilization, as well as
suspicion toward experts who wished to meddle with marriage. He was also
repeatedly critical of positive eugenics as a utopian pipe-dream that could
not take sufficient account of human depravity via original and actual sin.
For Maier, biblical preaching and traditional marriage were the solutions to
the breakdown of modern society, not the promises of an unrealistic eugenics movement.

Conclusion

The importance of Jonathan Edwards's treatise on original sin to American
theology has been noted above. It is therefore ironic that, given his emphasis on the universal scope of sin, the theologian's name would be invoked by
eugenicists to promote hereditary pride. Eugenics enthusiasts used Jonathan Edwards's name and family pedigree to contrast a eugenically healthy
lineage against numerous dysgenic lineages. Exemplary in this regard was
a sermon preached in the late 1920s for the Eugenics Sermon Contest
sponsored by the American Eugenics Society. In the text, the preacher told

---

61. W. Maier, *For Better Not for Worse*, 392.

a story "well known to all eugenists." He traced the heritage of Jonathan Edwards, whose marriage to Sarah Pierpont had resulted in descendants that included "twelve college presidents, two hundred and ninety-five college graduates, sixty-five college professors, one hundred clergymen, one hundred lawyers, seventy-five army officers, sixty doctors, eighty public officers, sixty prominent authors, and thirty judges." Not satisfied with this, he went on to regale his audience with further impressive lineal details: "In this line were governors, mayors, congressmen, United States senators and a vice-president of the United States."[62] To drive home the point of the importance of eugenic marriages, he drew a stark contrast between the Edwards lineage and the heritage of the infamously dysgenic families known as the Jukes and Kallikaks. Ironically, though Jonathan Edwards had strongly defended maintaining clear distinctions between natural and moral abilities, later students of heredity would invoke the name of Edwards in a process that repeatedly blurred and muddied such distinctions in prejudicial and harsh ways.

From about 1860 to 1940 North America saw a reassessment of human nature through the interplay of scientific and theological pronouncements on the ancient question of the psalmist: "What is man?"[63] Some would see new biological ideas from figures such as Darwin, Huxley, Spencer, and Weismann as occasioning an abandonment of ancient theological verities. In point of fact, Darwin himself agonized over the accusations of atheism that would be hurled at him, and the resentment his writings could create within his own family, as he anticipated the 1871 publication of *The Descent of Man*.[64] Yet others would see in the same findings new tools for apologetic restatement of traditional doctrines in a more scientific terminology.

The rhetorical and emotional force of preaching was often coupled with the authority and assumed objectivity of modern science in Gilded Age and Progressive Era social reform. When the "unfit" proved resistant to change by the social reformers' programs, a ready explanation presented itself: the new science of heredity. What if persons were not merely socially unredeemable because of environmental factors affecting individuals, but instead morally defective due to generations of bad breeding within whole families or populations? What if actual sins became over time hereditary moral defects, and acquired characteristics became instantiated into the very physiology of a germ plasm passed on by human procreation? Degeneration

---

62. Sermon marked #37, 1926, Box 11, American Eugenic Society Papers, American Philosophical Society, Philadelphia.

63. Psalm 8:1a, KJV.

64. See Desmond and Moore, *Darwin*, 568–81.

theory began to articulate a scientific rationale that carried with it a strong theological resonance. Degeneration theory blurred the categories of original and actual sin. As a scientific theory, it had little to say about Adam as a historical progenitor, but much to say about hereditary defectiveness. The biblical trope of "sins of the fathers to the third and fourth generation" took on not merely a homiletical, but also a scientific, cast. As a form of religious rhetoric, degeneration theory could attract reformers steeped in a Protestant religious ethos already wary of sin as both a hereditary and a behavioral problem.

The period between Darwin's *Origin* of 1859 and the Scopes Trial of 1925 witnessed a shift from a rather insular theology to a more public theology. While systematic theologians could operate comfortably in the arena of the seminary, for figures like Walter Maier the new tools of radio and slick magazine production and distribution in the 1920s gave impetus to a public engagements with trends of the times, including eugenics. The weight given to the idea of reversion or degeneration in science had a potent resonance within a theological tradition with its own notions of hereditary defect. Deep traditions rich with religious themes and rhetoric that held sway in the population of North America, and the intermixing of terminologies of theology, science, and social policy could at times render a toxic brew. Only a few theologically educated Protestants expressed anything like a principled and sustained public suspicion toward eugenic claims. The contours of this mixing process showed that the blending of theological and scientific categories with social reform policies could become tragically discriminatory toward the most vulnerable citizens in American society.

CONCLUSION

The Quest for Good Births

IN THESE PAGES I have argued that degeneration, a core concept of the eugenics movement, served as a key conceptual nexus between theological and scientific reflection on heredity among Protestant intellectuals and social reformers in the late nineteenth and early twentieth centuries. If this indeed is the historical reality, what are its implications for the twenty-first century?

That Americans are still seeking to promote "good births" may be discerned from numerous developments in the late twentieth and early twenty first centuries. Of course the desire for healthy children is perfectly natural and appropriate. Yet with increasing proficiency in the world of biotechnology and infertility intervention, several birth-related issues have begun to arise in scientific, legal, political, ethical and religious discourses. Contemporary science offers a proliferation of prenatal diagnostic tests. These include: amniocentesis, chorionic villus sampling, chromosome microarrays, biochemical marker screening, ultrasonography, nuchal translucency screening, and molecular testing. With such tools, medical specialists can inform parents of abnormalities or potential abnormalities of children while they are yet in the womb, most critically prior to twenty-four-weeks of gestation (i.e., when abortion remains a legal option in many states).

One growing legal consequence of this available information is the rise of "wrongful birth" lawsuits against medical professionals brought by parents of disabled children. Some of these parents claim medical personnel had either failed to inform or had under-informed them about the disability of their unborn child. They claim that with such information they would have terminated the pregnancy and forestalled much familial suffering. It is important to note that the physicians in question are not typically charged with causing the abnormalities or birth defects (such would be covered by traditional malpractice laws), but merely with not providing information about the (potential) disabilities of the offspring of those bringing suit.[1]

1. Whitney and Rosenbaum, "Recovery of Damages," 169. French physicians rose up in vocal protest against such lawsuits, and the French legislature was compelled

A 2011 study of the legal literature surrounding unwanted births indicated wrongful birth has been accepted as a cause of action in at least 29 jurisdictions within the USA. Only eight jurisdictions refused outright to accept such a cause of action. As of that year some fourteen jurisdictions had not taken a clear position on the issue.[2] Damages have been awarded to parents on various grounds, including the extraordinary medical and care-giving expenses of raising a disabled child. Some awards have even considered the care given beyond the age of majority. Courts are less united when it comes to the issue of damages for emotional distress.[3]

In one case that reached the South Carolina Supreme Court, the court resisted such a trend, concluding that: "South Carolina does not recognize a common law cause of action for wrongful life brought by or on behalf of a child born with a congenital defect because it is impossible to prove that being terminated by elective abortion, and thus never being born, is better than being born and living a life with disabilities."[4]

Not surprisingly, the proliferation of such lawsuits has received much attention in the bioethics literature. Philosophically, it would appear that implicit in such legal actions is the assumption that it would be better for at least some disabled persons not to exist at all than for them to exist and to suffer disabilities, and by extension, cause suffering (emotional and financial) among those charged with their care. Differences exist over whether the cause of legal action should be couched in terms of wrongful birth, wrongful disability, or wrongful existence. One understandable pocket of resistance to the cultural assumptions underlying such lawsuits regarding the value of disabled persons in society is the growing field of disability studies. Some scholars concerned about disability openly describe the use of prenatal testing for selecting the disabled out of the gene pool in terms explicitly warning of a new and troubling *eugenic* mindset. Others resist the drawing of connections to earlier eugenics, and find such lawsuits to be warranted and defensible on varying grounds.[5]

---

to place more protections in the law *vis a vis* wrongful birth actions in that country. See Lysaught, "Wrongful Life? The Strange Case of Nicholas Perruche," 9–11.

2. Whitney and Rosenbaum, "Recovery of Damages," 171–72.

3. Ibid., 173–88 (extraordinary expense analysis), 189–96 (emotional damage analysis).

4. Cited in Bostrom, "Willis v. Wu in the Supreme Court of South Carolina," 275.

5. Boss, *The Birth Lottery*, 1–326; Parens and Ash, *Prenatal Testing and Disability Rights*, 3–56; Childress, "Genetics, Disability, and Ethics," 157–79; Fergusson, "The One Who Smiled a Lot,"; Fox and Griffin, "Disability-Selective Abortion and the Americans with Disabilities Act," 845–905; Meehan, "Triumph of Eugenics in Prenatal Testing," 28–40; for the contrary view, cf. Strasser, "Yes Virginia, There can be Wrongful Life," 821–61; Mahowald, "Prenatal Testing for Selection Against Disabilities," 457–62.

Certainly very few intellectuals today embrace the label "eugenicist" openly and with enthusiasm. Historical awareness of the extreme and destructive social policies carried out under the banner of eugenics in the past inhibits open expressions of approval for controlling the procreation of others in a coercive manner. Efforts to control the procreation of inconvenient "others" may still be seen on occasion, such as the debate over whether judges may order women guilty of child abuse either to accept implantation of the birth control Norplant or to go to prison in the mid-1990s. Termination of pregnancies on the basis of Down Syndrome, Spina Bifida, and even the XYY karyotype (sometimes, albeit tenuously, associated with criminality) is not uncommon.[6] At the other end of the spectrum, some states, universities, and church denominations have apologized for their past support of eugenic sterilization activities.[7] Still, the yearning to control procreation, including the procreation of those who deviate from the ostensible "normal" mean continues to manifest itself in public policy debates of the twenty-first century.

Toward a Theology of Disability

In response to problematic attitudes toward persons with disabilities in the past and the present, a growing cadre of scholars is specializing in disability studies. Within that community are those who study disability from the perspective of theology, creating a genre of literature in the discipline known as the theology of disability. Such theology intersects with older, more established specializations such as historical theology, systematic theology, and practical theology.

One significant work exploring the history of the intersection of theological reflection and the experience of disability is *Disability in the Christian Tradition: A Reader*, edited by Brian Brock and John Swinton. The book is a blending of primary texts from the Christian traditions framed by introductory comments by several scholars in the field of the theology of disability. The editors include primary texts drawn from the Patristic Era, the Medieval period, the Reformation, and Modernity. Brian Brock situates disability studies within a fundamental reflection on theological anthropology. Here

6. Duster, "Persistence and Continuity," 228–29; Pioro, "Wrongful Birth Litigation," 1027–29.

7. Gormstyn, "Dean Apologizes for Medical School's Role in Sterilization Program," A12; Feist, "Davis Apologizes for State's Sterilization Program," A20; Kusmer, "Indiana Apologizes for Role in Eugenics," no page; Leslie and Chou, "Victims of NC forced sterilization program tell their stories," no page; "The United Methodist Church, An Apology for Support of Eugenics," no page.

he seeks to answer "the problem of the human." A significant task in the theology of disability is to interrogate and critique received rhetorical tropes that have had a tendency to demean, and indeed to dehumanize, those persons who depart from some ostensible norm of physical health or mental ability.[8] Focusing on a person's physical or intellectual abilities as a default way of measuring human worth is a distinctly modern problem.

Eugenics emerged at the peak of the influences of scientism and modernism. Many of the worst instances of such language have been illustrated across the foregoing chapters of this work. However during this same period the famed German theologian Dietrich Bonhoeffer, troubled by the rise of eugenics in Germany in the thirties, set forth a counter-cultural theological vision. He wrote: "There is no worthless life before God, because God holds life itself to be valuable. Because God is the Creator, Preserver, and Redeemer of life, even the poorest life before God becomes a valuable life."[9] Dietrich's brother Karl was a professor of psychiatry who used his position of authority to shield many disabled persons from the euthanasia campaigns carried out against the handicapped by the Nazi regime. Thus family dynamics also helped sensitize the famed theologian to the rising violence of the German government toward the mentally and physically disabled.[10]

Other recent works have focused on the role of disability studies in shaping discourse in theology. Myroslaw Tataryn and Maria Truchan-Tataryn have situated the discussion of disability within trinitarian theology. The trinity is an "inclusive community," that models the kind of wholeness and belonging that should lead to welcoming and including persons with disabilities in the life of the Christian fellowship.[11] Hans S. Reinders explores themes such as disability and divine providence, cosmic fairness, and critiques the ways Christians often speak of divine control or causation of disabilities in unhelpful ways. He seeks out first-person stories from families experiencing disability, and brings forth the voices of those for whom disability is an existential reality. In so doing Reinders clarifies ways in which glib and pietistic explanations of disability on one extreme, and rationalistic medicalized explanations on the other, should give way to a more sensitive discourse taking the concept of accessibility well beyond a mere changing of the architecture of the church building.[12]

---

8. Brock and Swinton, *Disability in the Christian Tradition*, 1–3.

9. Cited in Wannenwetsch, "'My Strength is Made Perfect in Weakness,'" 353.

10. Ibid., 360.

11. Tataryn and Truchan-Tataryn, *Discovering Trinity in Disability*, 23, 116–23.

12. Reinders, *Disability, Providence, and Ethics*, 1–52.

As historical and systematic theologies serve a role in creating space for an honest assessment of disability as a theological problem, practical theology seeks opportunities to demonstrate better patterns of inclusion and acceptance of persons with disabilities. Notable in this project is Thomas E. Reynolds, *Vulnerable Communion: A Theology of Disability and Hospitality*. Reynolds pursues a form of theological reflection with the goal of leading Christians to "think differently about disability and act differently toward people with disabilities," and foster "communities of abundant hospitality." This involves a reassessment of what it means to experience "wholeness" in the human community, and in the church. Reynolds argues that "wholeness is not the product of self-sufficiency or independence, but rather of the genuinely inclusive communion that results from sharing our humanity with one another in light of the grace of God."[13]

Overcoming the haunting spectre of the dehumanizing discourse of eugenics toward persons with disabilities will take special effort by communities of faith, and such a conversation begins in the religious academy. In this book I have sought to expose the hurtful language and invidious social policies disabled persons have had to endure in the past, as a way of urging upon the church an attitude of acceptance and embrace toward those whose disabling conditions ought not to lead to exclusion.

Yale Theologian Miroslav Volf offers helpful insights into the psychological process of rendering those who differ from us as "others," a process he calls "othering." Volf's main concern is inter-religious and inter-faith relations, but I believe his insights to be beneficial when considering a cogent response to the eugenics mindset. Much of the eugenics enthusiasm historically analyzed in this book utilized othering to an egregious degree. Whether elites were focused on problematic others such as the mentally defective, hereditary criminals, congenital inebriates, or allegedly inferior races, their rhetoric relied on a strong "us" versus "them" dichotomization of American society. Volf's analysis of a profoundly human tendency to keep others at a distance applies to our understanding of the mentality of the eugenics movement:

> That by which we differ from others is properly and exclusively our own, and it is in what is exclusively our own that our identity resides, we sometimes think. If we operate with such an exclusive notion of identity, we will watch carefully to make sure that no external elements enter our proper space so as to disturb the purity of our identity. Especially in situations of economic and political uncertainty and conflict, we will insist on pure identity.

13. Reynolds, *Vulnerable Communion*, 14, 18.

> If race matters to us, then we will want our "blood" to be pure,
> untainted by the "blood" of strangers. . . . If culture matters to us,
> then we will want our language and customs to be pure, cleansed
> of foreign words and foreign ways. This is the logic of purity. It
> attends the notion of identity, which rests on difference from
> the other. The consequences of the logic of purity in a pluralistic
> world are often deadly. We have to keep the other at bay, even
> by means of extreme violence, so as to avoid contamination.[14]

The widespread acceptance of eugenics, especially in the period
1907–1927, represented a failure of solidarity between Protestant elites and
those difficult "others" whose full participation in the human community of
society they denied. The theological notion of an inclusive posture toward
all human persons, both as created beings in the image of God, and as co-
participants in the tragedy of sin and brokenness, came into a doleful state
of neglect. Those who maintained a humane solidarity with those on the
margins of society were few, lonely and prophetic voices shouting in the
wilderness.

An Experiment in Rehumanization
in the Progressive Era

A century ago, one individual sought to shed light on the condition of prisoners, many of whom were regarded as hopeless cases, widely believed to be
criminals by their very nature, through the fateful machinations of remorse-
less heredity. This social reformer, in this instance a prison reformer in the
state of New York, cast a different vision. In the early twentieth century,
Thomas Mott Osborne (1859–1926) was the Commissioner on State Prison
Reform, appointed to that position by New York Governor Sulzner, on the
suggestion of Judge Riley, Superintendent of prisons. Osborne resolved,
with the input of many persons including former prisoners, to spend a week
as a prisoner at the Auburn Prison in Auburn, New York. This week of voluntary incarceration began on September 29, 1913. Writing about his plan,
Osborne explained the rationale for such a perilous undertaking:

> So the idea of some day entering prison and actually living the
> life of a convict first occurred to me more than three years ago.
> Talking with a friend, after his release from prison, concerning
> his own experience and the need of changes in the System, I
> brought forward the idea that it was impossible for those of us

14. Volf, "Living with the 'Other,'" 8–25.

on the outside to deal in full sympathy and understanding with
the man within the walls until we had come in close personal
contact with him, and had had something like a physical experience of similar conditions.[15]

Far from confirming conventional wisdom about the bestial inhumanity
of criminals, Osborne emerged ". . . with a new conception of the inherent nobility of human nature, a new belief in the power of men to respond
to the right conditions and the right appeal. I have come out with a new
sense of human brotherhood, a new faith in God."[16] Osborne entered with
a suspicion that much of what he had read in the literature of penology was
unreliable, given that this literature was written ". . . from such an outside
standpoint and with so little intelligent sympathy and vital understanding"
as to be without genuine value in getting at the truth.[17]

On Thursday, four days into his voluntary imprisonment, Osborne
held a conversation with a convict named Jack. They explored the idea of
creating greater opportunities for self-governance among the prisoners.
Jack's perspective differed greatly from eugenicists whose totalizing discourse held criminals to be congenitally defective degenerates. Osborne recorded Jack's point of view in these words: "Now if you trust a man, he'll try
and do what's right, sure he will. That is, most men well. Of course, there are
a few that won't. There are some dirty curs—degenerates—that will make
trouble, but there ain't so very many of those."[18]

After his release from a week of incarceration, Osborne penned some
summary remarks on the nature of prisoners and prison life. His inclusion
of considerations of religious themes fits well with the ethos of the time (and
the subject of this book), and indicates the implicit theological dimension
embedded in Protestant social reform efforts of the Progressive Era:

> As I leave the prison again, there ring in my ears the questions:
> What has happened? What does it all mean? It means just one
> thing—my friend—for it is you now, you individually, to whom
> I am speaking; it means that these prisoners are men—real
> men—your brethren—and mine. It means that if you treat them
> like beasts it will be hard for them to keep from degenerating
> into beasts. If you treat them like men you can help them to rise.
> It means that if you trust them they will show themselves wor-
> thy of trust. It means that if you place responsibility upon them

15. Osborne, *Within Prison Walls*, 4.
16. Ibid., 10.
17. Ibid., 16.
18. Ibid., 155.

they will rise to it. Perhaps some may think that I am leaving out of consideration the direct religious appeal that can be made to the prisoners. By no means. I have no intention of underrating the religious appeal. Under the old depressing conditions it is about the only appeal that can be made. But the religious appeal, to be really effective, must be based upon a treatment of the prisoner somewhat in accordance with the precepts of religion. Preaching a religion of brotherly love to convicts while you are treating them upon a basis of diabolical hatred is a discouraging performance.[19]

The student of the past must always exercise caution in criticizing persons of another era, or, conversely, of making them into angelic beings via hagiographical treatment. Yet we can hold them up to the scrutiny of the best impulses of their own time. One hundred years from now, our descendants may marvel at what they consider our own recalcitrant barbarisms, of which we are only now faintly conscious. Yet if they are responsible historians, they will locate voices of this era offering critiques of our own generation's lingering eugenic schemes. They will discern where we failed to acknowledge our own defective genomes, our own debilitating environments, and the mutually shared brokenness that is the universal human condition. They will also discover those moments when we discovered we truly needed one another, and spoke up for others amid our solidarity in the shared imperfection, and the shared dignity, of our humanity.

19. Ibid., 323–24.

Bibliography

Abbott, Lyman. *The Theology of an Evolutionist*. New York: Houghton Mifflin, 1897.

Aristotle. *Minor Works*. Translated by W. S. Hett. Loeb Classical Library. Cambridge, MA: Harvard University Press, 1963.

Asylum Projects. "Gallipolis Epileptic Hospital." http://www.asylumprojects.org/index.php?title=Gallipolis_Epileptic_Hospital.

Barker, Arthur A. "Sin: The Hydra-Headed Monster, IV: The Crime Phase." *The War Cry* (April 13, 1901), 1–5.

Barr, Martin W. *Mental Defectives: Their History, Treatment, and Training*. Philadelphia: P. Blakiston's Sons, 1904.

Barthelemy-Madaule, Madeleine. *Lamarck the Mythical Precursor: A Study of the Relations between Science and Ideology*. Cambridge, MA: Massachusetts Institute of Technology Press, 1982.

Bennetch, John Henry. "The Biography of Bibliotheca Sacra." In *Bibliotheca Sacra* 100 (1943) 8–12.

Black, Edwin. *War against the Weak: Eugenics and America's Campaign to Create a Master Race*. New York: Four Walls, Eight Windows, 2003.

Boardman, H. A. *The Scripture Doctrine of Original Sin Explained and Enforced in Two Discourses*. Philadelphia: William S. Martien, 1839.

Boddy, James M. "Brain Weight and Intellectual Development: Physical Variations of the Negro and the Anglo-Saxon Races." In *The Colored American Magazine* 9 (1905) 357–64.

———. "The Ethnology of the Japanese Race." In *The Colored American Magazine* 9 (1905) 577–85.

Boss, Judith A. *The Birth Lottery: Prenatal Diagnosis and Selective Abortion*. Chicago: Loyola University Press, 1993.

Bostrom, Barry A. "Willis v. Wu in the Supreme Court of South Carolina." In *Issues in Law and Medicine* 20 (2005) 275–78.

Bowers, Paul E. "The Recidivist." In *Journal of the American Institute of Criminal Law and Criminology* 5 (1914) 404–18.

Bowler, Peter J. *The Mendelian Revolution: The Emergence of Hereditarian Concepts in Modern Science and Society*. Baltimore, MD: Johns Hopkins University Press, 1989.

Bratt, James D. "Protestant Immigrants and the Protestant Mainstream." In *Minority Faiths and the American Protestant Mainstream*, edited by Jonathan D. Sarna, 110–35. Urbana, IL: University of Illinois Press, 1998.

Bray, Gerald. "Original Sin in Patristic Thought." In *Churchman* 108 (1994) 37–47.

Bremner, Robert H., and John Barnard. *Children and Youth in America: A Documentary History*. Cambridge, MA: Harvard University Press, 1971.

Brock, Brian, and John Swinton. *Disability in the Christian Tradition: A Reader*. Grand Rapids, MI: Eerdmans, 2012.

Brockway, Zebulon Reed. *Fifty Years of Prison Service: An Autobiography*. New York: Charities Publication Committee, 1912.

Bruinius, Harry. *Better for All the World: The Secret History of Forced Sterilization and America's Quest for Racial Purity*. New York: Knopf, 2006.

Bucke, Emory Stevens, ed. *The History of American Methodism*, Volume 3. Nashville, TN: Abingdon, 1964.

Burbank, Luther. *The Training of the Human Plant*. New York: Windham, 1907.

Burger, William. *American Crime and Punishment: The Religious Origins of American Criminology*. Buchanan, MI: Vande Vere, 1993.

Burkhardt, Richard W. Jr. *The Spirit of System: Lamarck and Evolutionary Biology*. Cambridge, MA: Harvard University Press, 1977.

Butler, Amos W. "At the End of the Century." In *Indiana Bulletin of Charities and Correction* (1915) 90–99.

———. "The Board of State Charities and the People." In *Indiana Bulletin of Charities and Correction* (1908) 1–5.

———. "President's Address: The Burden of Feeble-mindedness." In *Indiana Bulletin of Charities and Correction* (1911) 297–308.

Bynum, W. F. "Alcoholism and Degeneration in Nineteenth Century European Medicine and Psychiatry." In *British Journal of Addiction* 79 (1984) 59–70.

Carey, Allison C. *On the Margins of Citizenship: Intellectual Disability and Civil Rights in Twentieth-Century America*. Philadelphia: Temple University Press, 2009.

Carlson, Elof A. *The Unfit: History of a Bad Idea*. Cold Spring Harbor, NY: Cold Spring Harbor Laboratory, 2001.

Carson, James C. "Prevention of Feeble-mindedness from a Moral and Legal Standpoint: Report of Committee." In *Proceedings of the National Conference of Charities and Correction*, edited by Isabel C. Barrows, 294–303. Boston: George H. Ellis, 1898.

Cashdollar, Charles D. *The Transformation of Theology, 1830–1890: Positivism and Protestant Thought in Britain and America*. Princeton, NJ: Princeton University Press, 1989.

Chamberlin, J. Edward, and Sander L. Gilman, eds. *Degeneration: The Dark Side of Progress*. New York: Columbia University Press, 1985.

Childress, Kelly D. "Genetics, Disability, and Ethics: Could Applied Technologies lead to a New Eugenics?" In *Journal of Women & Religion* 20 (2002) 157–78.

Clark, Charles M. *The Prevention of Racial Deterioration and Degeneracy*. Ohio Board of Administration, Publication 15. Mansfield, OH: Ohio State Reformatory, 1920.

Coffin, Jean-Christophe. "Heredity, Milieu and Sin: The Works of Bénédict Augustin Morel (1809–1873)." In *A Cultural History of Heredity 2: 18th and 19th Centuries*, edited by Hans-Jorg Rheinberger and Staffan Muller-Wille, 153–64. Preprint 247, submitted 2003. http://pubman.mpiwg-berlin.mpg.de/pubman/item/escidoc:644412:3/component/escidoc:644410/P247.pdf#page=154.

———. "Le Théme de la Degénerescence de la Race autour de 1860." In *History of European Ideas* 15 (1992) 27–32.

Cohen, Adam. *Imbeciles*. New York: Penguin, 2016.

Conn, H. J. "Professor Herbert William Conn and the Founding of the Society." In *Bacteriology Reviews* 12 (1948) 275–96.

Conn, Herbert William. "Eugenics versus Social Heredity." In *The Methodist Review* 95 (1913) 708–12.

———. *Social Heredity and Social Evolution: The Other Side of Eugenics*. New York: Abingdon, 1914.

Cook, Raymond A. *Thomas Dixon*. New York: Twayne, 1974.

Cope, Edward Drinker. *The Origin of the Fittest: Essays on Evolution*. New York: Appleton, 1887.

Cravens, Hamilton. *Triumph of Evolution: American Scientists and the Heredity-Environment Controversy, 1900–1914*. Philadelphia: University of Pennsylvania Press, 1978.

Currier, Albert H. "Crime in the United States: Reforms Demanded." In *Bibliotheca Sacra* 68 (1911) 61–93.

Davenport, Charles Benedict. *Eugenics as a Religion*. Cold Spring Harbor, NY: Eugenic Record Office, 1916.

———. *Heredity in Relation to Eugenics*. New York: Henry Holt, 1911.

Davenport, Charles B., and David F. Weeks. *A First Study of Inheritance in Epilepsy*. Cold Spring Harbor, NY: Eugenic Record Office, 1911.

Davis, Edith Smith. *A Compendium of Temperance Truth: Largely Contributed by the Counselors of the Department of Scientific Temperance Investigation and of Scientific Temperance Instruction of the World's and National Woman's Christian Temperance Union*. Milwaukee: Advocate, 1916.

Degler, Carl. *In Search of Human Nature: The Decline and Revival of Darwinism in American Social Thought*. New York: Oxford University Press, 1991.

Desmond, Adrian, and James Moore. *Darwin: The Life of a Tormented Evolutionist*. New York: W. W. Norton, 1991.

Deutsch, Nathaniel. *Inventing America's "Worst" Family: Eugenics, Islam, and the Fall and Rise of the Tribe of Ishmael*. Oakland, CA: University of California Press, 2009.

Dix, Dorothea. "'I Tell What I have Seen': The Reports of Asylum Reformer Dorothea Dix." In *American Journal of Public Health* 96 (2006) 622–24.

Dixon, Thomas. *The Leopard's Spots: A Romance of the White Man's Burden*. New York: Doubleday, 1905.

Dowbiggin, Ian Robert. *Keeping America Sane: Psychiatry and Eugenics in the United States and Canada, 1880–1940*. Ithaca, NY: Cornell University Press, 1997.

Drähms, August. *The Criminal: His Personnel and Environment, A Scientific Study*. New York: Macmillan, 1900.

Dunlap, Mary J. "Progress in the Care of the Feebleminded and Epileptics." In *Proceedings of the National Conference of Charities and Correction*, edited by Isabel C. Barrows, 255–59. Boston: George H. Ellis,1900.

Durst, Dennis L. "The Future in 1916: Madison Grant and *The Passing of the Great Race*." In *Nomocracy in Politics* (2016), no pages. https://nomocracyinpolitics.com/2016/01/21/the-future-in-1916-madison-grant-and-the-passing-of-the-great-race-by-dennis-l-durst/.

Dugdale, Richard L. *"The Jukes": A Study in Crime, Pauperism, Disease and Heredity: And Further Studies of Criminals*. New York: Putnam's, 1877.

Duster, Troy. "Persistence and Continuity in Human Genetics and Social Stratification." In *Genetics: Issues of Social Justice*, edited by Ted Peters, 218–37. Cleveland, OH: Pilgrim, 1998.

Eadie, M. J., , and F. J. E. Vajda. *Antiepileptic Drugs: Pharmacology and Therapeutics*. Berlin: Springer-Verlag, 1999.

Eames, Blanche. *Principles of Eugenics: A Practical Treatise*. New York: Moffat, Yard, 1914.

Edwards, Jonathan. "The Great Christian Doctrine of Original Sin Defended." In *A Jonathan Edwards Reader*, edited by J.E. Smith, et al., 223–43. New Haven, CT: Yale University Press, 1995.

Edwards, Justin D. *Gothic Passages: Racial Ambiguity and the American Gothic*. Iowa City, IA: University of Iowa Press, 2003.

Ellis, Havelock. *Little Essays of Love and Virtue*. New York: George H. Doran, 1922.

Ellwood, Charles A. *The Reconstruction of Religion: A Sociological View*. New York: Macmillan, 1923.

Engs, Ruth Clifford. *The Eugenics Movement: An Encyclopedia*. Westport, CT: Greenwood, 2005.

Estabrook, Arthur H., and Charles B. Davenport. *The Nam Family: A Study in Cacogenics*. Cold Spring Harbor, NY: Eugenics Record Office, 1912.

Evans, Richard J. *The Third Reich at War*. New York: Penguin, 2010.

Fabian, Ann. *The Skull Collectors: Race, Science, and America's Unburied Dead*. Chicago: University of Chicago Press, 2010.

Feist, Paul. "Davis Apologizes for State's Sterilization Program: Those with Hereditary Flaws Were Victims." In *San Francisco Chronicle*, March 12, 2003.

Ferguson, Philip M. *Abandoned to Their Fate: Social Policy and Practice toward Severely Retarded People in America, 1820–1920*. Philadelphia: Temple University Press, 1994.

Fergusson, Andrew. "The One Who Smiles a Lot." In *Center for Bioethics & Human Dignity*, February 3, 2006, no pages. https://cbhd.org/content/one-who-smiles-lot.

Finlayson, Anna Wendt. *The Dack Family: A Study in Hereditary Lack of Emotional Control*. Cold Spring Harbor, NY: Cold Spring Harbor Laboratory, 1916.

Fisher, James T. *Catholics in America*. Religion in American Life. New York: Oxford University Press, 2000.

Fox, Dov, and Christopher L. Griffin Jr. "Disability-Selective Abortion and the Americans with Disabilities Act." In *Utah Law Review* (2009) 845–905.

French, J. M. "The Prognosis of Inebriety." In *The Quarterly Journal of Inebriety* 20 (1898) 17–30.

Friedlander, Walter J. *The History of Modern Epilepsy: The Beginnings, 1865–1914*. New York: Praeger, 2001.

Galton, Francis. *Hereditary Genius: An Inquiry into Its Laws and Consequences*. Gloucester: Peter Smith, 1972.

———. *Inquiries into Human Faculty and Its Development*. New York: E.P. Dutton, 1908.

Gamwell, Lynn, and Nancy Tomes. *Madness in America: Cultural and Medical Perceptions of Mental Illness before 1914*. Ithaca, NY: Cornell University Press, 1995.

Gates Jr., Henry Louis. "Harlem on Our Minds." In *Critical Inquiry* 24 (1997) 1–12.

Giele, Janet Zollinger. *Two Paths to Women's Equality: Temperance, Suffrage, and the Origins of Modern Feminism.* New York: Twayne, 1995.

Gillin, John Lewis. *Criminology and Penology.* New York: Century, 1926.

―――. *Poverty and Dependency: Their Relief and Prevention.* New York: Century, 1928.

―――. *Social Pathology.* New York: Century, 1933.

―――. *Social Problems.* New York: Century, 1928.

―――. "The Study of Social Religion and Social Reconstruction." In *Religious Education* 15 (1920) 103–105.

Gilman, Stuart C. "Degeneracy and Race in the Nineteenth Century: The Impact of Clinical Medicine." In *Journal of Ethnic Studies* 10 (1983) 27–50.

Gobineau, Arthur de. *The Inequality of Human Races.* New York: Howard Fertig, 1967.

Goddard, Henry Herbert. *Feeble-Mindedness: Its Causes and Consequences.* New York: Macmillan, 1923.

―――. *The Kallikak Family.* New York: Macmillan, 1912.

―――. *The Kallikaks: A Study in the Heredity of Feeble-Mindedness.* New York: Macmillan, 1912.

Gordon, Elizabeth Putnam. *Women Torchbearers: The Story of the Woman's Christian Temperance Union.* Evanston, IL: National Women's Christian Temperance Union, 1924.

Gormstyn, Alice. "Dean Apologizes for Medical School's Role in Sterilization Program." In *Chronicle of Higher Education*, November 14, 2003, A12.

Gould, Stephen Jay. *The Mismeasure of Man.* New York: W. W. Norton, 1996.

Green, Sanford M. *Crime: Its Nature, Causes, Treatment, and Prevention.* Philadelphia: J.B. Lippincott, 1889.

Gritsch, Eric W. *A History of Lutheranism.* Minneapolis, MN: Fortress, 2002.

Gundry, Richard. "On the Care and Treatment of the Insane." In *Proceedings of the National Conference of Charities and Correction,* edited by Isabel C. Barrows, 253–75. Boston: George H. Ellis, 1890.

Hall, Francis J. *Creation and Man.* New York: Longmans Green, 1912.

―――. *Evolution and the Fall.* New York: Longmans Green, 1910.

Hamilton, James E. "The Church as a Universal Reform Society: The Social Vision of Asa Mahan." In *Wesleyan Theological Journal* 25 (1990) 42–56.

Hamilton, Virginia, ed. *The Writings of W. E. B. DuBois.* New York: Thomas W. Crowell, 1975.

Handlin, Oscar. *The Americans: A New History of the People of the United States.* Boston: Little Brown, 1963.

Handy, Robert T. *Undermined Establishment: Church-State Relations in America.* Princeton, NJ: Princeton University Press, 1991.

Hannaford, Ivan. *Race: The History of an Idea in the West.* Baltimore, MD: Johns Hopkins University Press, 1996.

Haroutunian, Joseph. *Piety versus Moralism: The Passing of the New England Theology.* Hamden, CT: Archon, 1964.

Harrell, David Edward Jr. "Bipolar Protestantism: The Straight and Narrow Ways." In *Re-Forming the Center: American Protestantism, 1900 to the Present,* edited by Douglas Jacobsen and William Vance Trollinger Jr., 15–30. Grand Rapids, MI: Eerdmans, 1998.

Hasian, Marouf Arif Jr. *The Rhetoric of Eugenics in Anglo-American Thought.* Athens: University of Georgia Press, 1996.

Hassin, Ran and Yaacov Trope. "Facing Faces: Studies on the Cognitive Aspects of Physiognomy." In *Journal of Personality and Social Psychology* 78 (2000) 837–52.

Hilts, Victor L. "Obeying the Laws of Hereditary Descent: Phrenological Views on Inheritance and Eugenics." In *Journal of the History of the Behavioral Sciences* 18 (1982) 62–77.

Holifield, E. Brooks. *Theology in America: Christian Thought from the Age of the Puritans to the Civil War*. New Haven, CT: Yale University Press, 2003.

Holloway, Jonathan Scott. "Harlem Renaissance Scholars Debate the Route to Racial Progress." In *The Journal of Blacks in Higher Education* 8 (1995) 60–63.

Horsman, Reginald. *Josiah Nott of Mobile: Southerner, Physician, and Racial Theorist*. Baton Rouge, LA: Louisiana State University Press, 1987.

Hunt, Mary Hannah. *A History of the First Decade of the Department of Scientific Temperance*. Boston: Washington, 1892.

Jacobs, J. "Care of the Mentally Retarded." In *Canadian Family Physician* 25 (1979) 1343–48.

James, William. *The Varieties of Religious Experience: A Study in Human Nature*. New York: Longmans, Green, 1902.

Johnson, Alexander. "The Segregation of Defectives: Report of the Committee on Colonies for Segregation of Defectives." In *Official Proceedings of the Annual Meeting of the Conference of Charities and Correction*, edited by Isabel C. Barrows, 245–49. Columbus, OH: Fred J. Heer, 1903.

Johnson, Miriam McNown, and Rita Rhodes. "Institutionalization: A Theory of Human Behavior and the Social Environment." In *Advances in Social Work* 8 (2007) 219–36.

Kellogg, John Harvey. "The Degeneration of the Negro." In *Good Health* 43 (1908) 588.

———. "Deterioration in Great Britain." In *Good Health* 39 (1904) 332.

———. "Mendel's Law of Heredity and Race Degeneration." In *Good Health* 45 (1910) 735–37.

———. "The Race Growing Old." In *Good Health* 41 (1906) 668.

———. "Recent Facts Regarding the Growing Prevalence of Race Degeneracy." In *Good Health* 48 (1913) 123.

Kerr, K. Austin. *Organized for Prohibition: A New History of the Anti-Saloon League*. New Haven, CT: Yale University Press, 1985.

"Kevin's Pyramids: A Stained Glass Collection for the Benefit of Epilepsy Therapies and Awareness," no pages. http://www.kevinspyramids.com/the-film.html.

Kevles, Daniel J. *In the Name of Eugenics: Genetics and the Uses of Human Heredity*. Cambridge, MA: Harvard University Press, 1995.

Kidd, Colin. *The Forging of Races: Race and Scripture in the Protestant Atlantic World, 1600–2000*. New York: Cambridge University Press, 2006.

Kline, Wendy. *Building a Better Race*. Berkeley, CA: University of California Press, 2001.

Knight, George. "The State's Duty toward Epileptics." In *Proceedings of the National Conference of Charities and Correction*, edited by Isabel C. Barrows, 298–302. Boston: George H. Ellis, 1886.

Koonz, Claudia. "Ethical Dilemmas and Nazi Eugenics: Single-Issue Dissent in Religious Contexts." In *Journal of Modern European History Supplement* 64 (1992) S8–S31.

Kühl, Stefan. *The Nazi Connection: Eugenics, American Racism, and German National Socialism*. New York: Oxford University Press, 1994.

Kusmer, Ken. "Indiana Apologizes for Role in Eugenics." In *Washington Post* April 13, 2007, no pages. http://www.washingtonpost.com/wp-dyn/content/article/2007/04/13/AR2007041300259_pf.html.

Landman, J. H. *Human Sterilization: The History of the Sexual Sterilization Movement.* New York: Macmillan, 1932.

Largent, Mark A. *Breeding Contempt: The History of Coerced Sterilization in the United States.* New Brunswick, NJ: Rutgers University Press, 2008.

Laughlin, Harry H. *Eugenical Sterilization in the United States.* Chicago: Psychopathic Laboratory of the Municipal Court of Chicago, 1922.

Leach, Mark W. "The Reckless, Profitable Elimination of Down Syndrome." In *The Witherspoon Institute*, November 10, 2011, no pages. http://www.thepublicdiscourse.com/2011/11/4240/.

Lecklider, Aaron. *Inventing the Egghead: The Battle over Brainpower in American Culture.* Philadelphia: University of Pennsylvania Press, 2013.

Leonard, Thomas C. *Illiberal Reformers.* Princeton, NJ: Princeton University Press, 2016.

Leslie, Laura, and Renee Chou. "Victims of NC Forced Sterilization Program Tell Their Stories." In *The Associated Press,* June 22, 2011, no pages. http://www.wral.com/news/state/nccapitol/story/9761898/.

Letchworth, William Pryor. *Care and Treatment of Epileptics.* New York: G.P. Putnam's Sons, 1900.

———. "The Care of Epileptics." In *Proceedings of the National Conference of Charities and Correction,* edited by Isabel C. Barrows, 199–206. Boston: George H. Ellis, 1896.

Liégeois, Alex. "La Théologie et L'éthique sous-jacentes à la Psychiatrie de B. A. Morel." In *Ephemerides Theologicae Lovaniensis* 65 (1989) 330–57.

Lippy, Charles H. *Religious Periodicals of the United States: Academic and Scholarly Journals.* New York: Greenwood, 1986.

Lombardo, Paul A. *Three Generations, No Imbeciles: Eugenics, the Supreme Court, and Buck vs. Bell.* Baltimore, MD: Johns Hopkins University Press, 2010.

Lombroso, Cesare. *Criminal Man.* Edited and translated by Mary Gibson and Nicole Hahn Rafter. Durham, NC: Duke University Press, 2006.

Ludmerer, Kenneth M. *Genetics and American Society.* Baltimore, MD: John Hopkins University Press, 1972.

Lysaught, M. Therese. "Wrongful Life?: The Strange Case of Nicholas Perruche." In *Commonweal* (2002) 9–11.

McClymond, Michael J. "John Humphrey Noyes, the Oneida Community, and Male Continence." In *Religions of the United States in Practice*, edited by Colleen McDannell, 1:218–33. Princeton Readings in Religions. Princeton, NJ: Princeton University Press, 2001.

MacDonald, Arthur. *Abnormal Man, Being Essays on Education and Crime and Related Subjects, with Digests of Literature and a Bibliography.* Washington DC: Government Printing Office, 1893.

———. *Juvenile Crime and Reformation Including Stigmata of Degeneration: Being a Hearing on the Bill to Establish a Laboratory for the Study of the Criminal, Pauper, and Defective Classes.* Washington DC: Government Printing Office, 1908.

Maffly-Kipp, Laurie F. "Mapping the World, Mapping the Race: The Negro Race History, 1874–1915." In *Church History* 64 (1995) 610–26.

Mahowald, Mary B. "Prenatal Testing for Selection against Disabilities." In *Cambridge Quarterly of Healthcare Ethics* 16 (2007) 457–62.

Maier, Paul L. *A Man Spoke, The World Listened: The Story of Walter A. Maier.* St. Louis, MO: Concordia, 1980.

Maier, Walter A. "Away with Our 'Splendid Isolation!'" In *Walther League Messenger* 38 (1929) 72–73.

———. *Christ for the Nation!: The Radio Messages Broadcast in the Third Lutheran Hour.* St. Louis, MO: Concordia, 1936.

———. *The Cross from Coast to Coast: Radio Messages Broadcast in the Fifth Lutheran Hour.* St. Louis, MO: Concordia, 1938.

———. *For Better Not for Worse: A Manual of Christian Matrimony.* 3rd ed. St. Louis, MO: Concordia, 1939.

———. *Fourth Lutheran Hour: Winged Words for Christ.* St. Louis, MO: Concordia, 1937.

———. "Here Comes the Pride!" In *Walther League Messenger* 35 (1927) 536–37.

———. *The Lutheran Hour: Winged Words to Modern America Broadcast in the Coast-to-Coast Radio Crusade for Christ.* St. Louis, MO: Concordia, 1931.

———. "Must Marriage Go Eugenic?" In *Walther League Messenger* 42 (1934) 328–29.

———. *The Radio for Christ: Radio Messages Broadcast in the Sixth Lutheran Hour.* St. Louis, MO: Concordia, 1939.

———. "Worse Than Plunging Parachutes." In *Walther League Messenger* 38 (1930) 266–67, 319.

Mathews, Shailer. *Contributions of Science to Religion.* New York: D. Appleton, 1924.

———. *New Faith for Old: An Autobiography.* New York: Macmillan, 1936.

McCulloch, Oscar C. *The Tribe of Ishmael: A Study in Social Degradation.* Indianapolis, IN: Charity Organization Society, 1888.

Meehan, Mary. "The Triumph of Eugenics in Prenatal Testing." In *Human Life Review* 35 (2009) 28–40.

Mehler, Barry Alan. "A History of the American Eugenics Society, 1821–1940." PhD diss., University of Illinois, 1988.

Miley, John. *Systematic Theology.* New York: Eaton & Mains, 1892.

Milum, J. Parton. *Evolution and the Spirit of Man.* London: Epworth, 1928.

Montagu, Ashley. *Race and IQ.* Expanded Edition. New York: Oxford University Press, 1999.

Moore, James R. *The Post-Darwinian Controversies: A Study of the Protestant Struggle to Come to Terms with Darwin in Great Britain and America.* New York: Cambridge University Press, 1979.

Morel, Bénédict A. *Traité des Dégénérescences Physiques, Intellectuelles et Morales de l'Espèce Humaine.* New York: Arno, 1976.

Morison, William James. "George Frederick Wright: In Defense of Darwinism and Fundamentalism." PhD diss., Vanderbilt University, 1971.

Morone, James A. *Hellfire Nation: The Politics of Sin in American History.* New Haven, CT: Yale University Press, 2003.

Mosby, Thomas Speed. *Causes and Cures of Crime.* St. Louis, MO: C.V. Mosby, 1913.

Nelson, E. Clifford. *The Lutherans in North America.* Philadelphia: Fortress, 1980.

Ngai, Mae M. *Impossible Subjects: Illegal Aliens and the Making of Modern America.* Princeton, NJ: Princeton University Press, 2004.

Noll, Mark A. *America's God: From Jonathan Edwards to Abraham Lincoln*. New York: Oxford University Press, 2002.

———. "The Bible, Minority Faiths, and the American Protestant Mainstream, 1860–1925." In *Minority Faiths and the American Protestant Mainstream*, edited by Jonathan D. Sarna, 191–231. Chicago: University of Illinois Press, 1998.

Norwood, Frederick A. *The Story of American Methodism*. Nashville, TN: Abingdon, 1975.

Numbers, Ronald L. *The Creationists*. New York: Alfred A. Knopf, 1992.

———. "George Frederick Wright: From Christian Darwinist to Fundamentalist." *Isis* 79 (1988) 624–45.

Nye, Robert A. "Heredity or Milieu: The Foundations of Modern European Criminological Theory." In *Isis* 67 (1976) 335–55.

Okrent, Daniel. *Last Call: The Rise and Fall of Prohibition*. New York: Scribner, 2010.

Orebaugh, David A. *Crime Degeneracy and Immigration: Their Interrelations and Interreactions*. Boston: Gorham, 1929.

Osborne, Thomas Mott. *Within Prison Walls: Being a Narrative of Personal Experience During a Week of Voluntary Confinement in the State Prison at Auburn, New York*. New York: Appleton, 1914.

Parker, Alison M. *Purifying America: Women, Cultural Reform, and Pro-censorship Activism*. Urbana, IL: University of Illinois Press, 1997.

Paul, Diane B. *Controlling Human Heredity 1865 to the Present*. Atlantic Highlands, NJ: Humanities, 1995.

Pauly, Philip J. *Biologists and the Promise of American Life: From Meriwether Lewis to Alfred Kinsey*. Princeton, NJ: Princeton University Press, 2000.

Peters, John L. *Christian Perfection and American Methodism*. New York: Abingdon, 1956.

Pick, Daniel. *Faces of Degeneration: A European Disorder, c. 1848–c. 1918*. New York: Cambridge University Press, 1989.

Pioro, Mark. "Wrongful Birth Litigation and Prenatal Screening." In *Canadian Medical Association Journal* 179 10 (2008) 1027–30.

Piper, John F., Jr. *Robert E. Speer: Prophet of the American Church*. Louisville, KY: Geneva, 2000.

Pisciotta, Alexander W. *Benevolent Repression: Social Control and the American Reformatory-Prison Movement*. New York: New York University Press, 1994.

Polglase, William A. "The Evolution of the Care of the Feeble-Minded and Epileptic in the Past Century." In *Proceedings of the National Conference of Charities and Correction*, edited by Isabel C. Barrows, 186–90. Boston: George H. Ellis, 1901.

"Preachers and Eugenics." Unsigned Editorial. In *American Breeders Magazine* 4 1 (1913) 62–64.

Rafter, Nichole Hahn. *Creating Born Criminals: Biological Theories of Crime & Eugenics*. Urbana, IL: University of Illinois Press, 1997.

———. *Origins of Criminology: A Reader*. New York: Routledge-Cavendish, 2009.

———. *White Trash*. Boston: Northeastern University Press, 1988.

Ransom, Julius B. "The Prison Physician and His Work: Health and Disease in Prison." In *Penal and Reformatory Institutions*. Edited by Charles R. Henderson, 261–90. New York: Charities Publication Committee, 1910.

Redfield, Casper Lavater. *Control of Heredity: A Study of the Genesis of Evolution and Degeneracy*. Chicago: Monarch, 1903.

Reed, James. *The Birth Control Movement and American Society: From Private Vice to Public Virtue*. Princeton, NJ: Princeton University Press, 1983.

Reinders, Hans S. *Disability, Providence, and Ethics: Bridging Gaps, Transforming Lives*. Waco, TX: Baylor University Press, 2014.

Report of the National Woman's Christian Temperance Union Twenty-Second Annual Meeting. Chicago: Woman's Temperance Publishing Association, 1895.

Reuber, Markus. *Epilepsy Explained: A Book for People Who Want to Know More*. New York: Oxford University Press, 2009.

"Rev. Oscar C. McCulloch Papers." Rare Books and Manuscripts Division, Indiana State Library. Updated 11/04/2015. http://www.in.gov/library/finding-aid/L363_McCulloch_Rev_Oscar_C_Papers.pdf.

Reynolds, Thomas E. *Vulnerable Communion: A Theology of Disability and Hospitality*. Grand Rapids, MI: Brazos, 2008.

Rimke, Heidi and Alan Hunt. "From Sinners to Degenerates: The Medicalization of Morality in the Nineteenth Century." In *History of the Human Sciences* 15 (2002) 59–88.

Rogers, A.C., and Maud A. Merrill. *Dwellers in The Vale of Siddem: A True Story of the Social Aspect of Feeble-Mindedness*. Boston: Richard G. Badger, 1919.

Ronk, Albert T. *History of the Brethren Church: Its Life, Thought, Mission*. Ashland, OH: Brethren, 1968.

Roosevelt, Theodore. *The Foes of our own Household*. New York: George H. Doran, 1917.

Rosen, Christine. *Preaching Eugenics*. New York: Oxford University Press, 2004.

Rosenberg, Charles E. *No Other Gods*. Baltimore, MD: Johns Hopkins University Press, 1976.

Russell, Ira. "A Paper on the Care of Epileptics." In *Proceedings of the Eighth Annual Conference of Charities and Correction*, edited by F.B. Sanborn, 325–30. Boston: A. Williams, 1881.

Savage, Minot J. *Men and Women*. Boston: American Unitarian Association, 1902.

Scott, Joseph F. "Prison and Police Administration Report of the Committee." In *Proceedings of the National Conference of Charities and Correction*, edited by Alexander Johnson, 86–91. Indianapolis, IN: William B. Burford, 1907.

Seitler, Dana. *Atavistic Tendencies: The Culture of Science in American Modernity*. Minneapolis, MN: University of Minnesota Press, 2008.

Shannon, T. W. *Nature's Secrets Revealed: Scientific Knowledge of the Laws of Sex Life and Heredity or Eugenics*. Marietta, OH: S.A. Mullikin, 1917.

Shorvon, Simon D. "Drug Treatment of Epilepsy in the Century of the ILAE: The First 50 years, 1909–1958." In *Epilepsia* 50 3 (2009) 69–92.

———. "The Early History of Epilepsia, the Journal of the International League against Epilepsy, and Its Echoes Today." In *Epilepsia* 48 1 (2007) 1–14.

Showalter, Nathan D. *The End of a Crusade: The Student Volunteer Movement for Foreign Missions and Great War*. Lanham, MD: Scarecrow, 1998.

Smith, David Livingstone. *Less than Human: Why we Demean, Enslave, and Exterminate Others*. New York: St. Martins, 2011.

Smith, J.S. "Marriage, Sterilization and Commitment Laws Aimed at Decreasing Mental Deficiency." In *Journal of the American Institute of Criminal Law and Criminology* 5 (1914) 364–70.

Smith, John David, ed. *Anti-Black Thought 1863–1925: "The Negro Problem."* Volume 6, *The Biblical and "Scientific" Defense of Slavery*, by John David Smith. New York: Garland, 1993.

Smith, Timothy L. *Revivalism & Social Reform: American Protestantism on the Eve of the Civil War*. Baltimore, MD: Johns Hopkins University Press, 1980.

Smyth, Newman. *Constructive Natural Theology*. New York: Charles Scribner's Sons, 1913.

Special Olympics International. "Rosa's Law." no pages. http://www.specialolympics.org/Regions/north-america/News-and-Stories/Stories/Rosa-s-Law.aspx.

Speer, Robert E. "Foreign Missions: An Experiment in Eugenics." In *Union Seminary Review* 56 (1944) 39–50.

———. *Race and Race Relations: A Christian View of Human Contacts*. New York: Fleming H. Revell, 1924.

Spratling, William P. *Epilepsy and Its Treatment*. New York: W.B. Saunders, 1904.

Spurgeon, C. H. *Treasury of David: Containing an Original Exposition of the Book of Psalms; a Collection of Illustrative Extracts from the Whole Range of Literature; A Series of Homiletical Hints upon Almost Every Verse; and Lists of Writers upon Each Psalm*. New York: Funk & Wagnalls, 1892.

Stoddard, Lothrop. *The Rising Tide of Color against White World Supremacy*. New York: Scribner, 1920.

Stokes, Mason. *The Color of Sex: Whiteness, Heterosexuality, and the Fictions of White Supremacy*. Durham, NC: Duke University Press, 2001.

Strasser, Mark. "Yes, Virginia, There Can Be Wrongful Life: On Consistency, Public Policy and the Birth-Related Torts." In *Georgetown Journal of Gender and Law* 4 (2003) 821–61.

Strode, Aubrey E. "Sterilization of Defectives." In *Virginia Law Review* 11 (1925) 296–301.

Strong, Augustus Hopkins. "Degeneration." In *Miscellanies* 2:110–28. Philadelphia: Griffith & Rowland, 1912.

Strong, Josiah. *Our Country: Its Possible Future and Its Present Crisis*. New York: Baker and Taylor, 1885.

Sumner, Walter Taylor. "The Health Certificate—A Safeguard Against Vicious Selection in Marriage." In *Proceedings of the First National Conference on Race Betterment*, edited by Emily F. Robbins, 509–13. Battle Creek, MI: Gage, 1914.

Sutherland, Edwin H. *Criminology*. Philadelphia: J. P. Lippincott, 1924.

Sutherland, Edwin H., et al. "The Theory of Differential Association." In *Deviance: A Symbolic Interactionist Approach*, edited by Nancy J. Herman, 64–68. Lanham, MD: General Hall, 1995.

Szasz, Ferenc Morton. *The Divided Mind of Protestant America, 1880–1930*. Huntsville, AL: University of Alabama Press, 1982.

Talbot, Eugene S. *Degeneracy: Its Causes, Signs, and Results*. New York: Charles Scribner's Sons, 1898.

Tataryn, Myroslaw, and Maria Truchan-Tataryn. *Discovering Trinity in Disability: A Theology for Embracing Difference*. Maryknoll, NY: Orbis, 2013.

Taylor, Carol M. "W. E. B. DuBois's Challenge to Scientific Racism." In *Journal of Black Studies* 11 (1981) 449–60.

Teats, Mary E. *The Way of God in Marriage: A Series of Essays upon Gospel and Scientific Purity*. Spotswood, NJ: Physical Culture, 1906.

Townsend, Luther T. *Collapse of Evolution*. Louisville, KY: Pentecostal, 1921.

Tracy, Sarah W. *Alcoholism in America: From Reconstruction to Prohibition*. Baltimore, MD: Johns Hopkins University Press, 2005.

Trent, James. *Inventing the Feeble Mind: A History of Mental Retardation in the United States*. Berkeley, CA: University of California, 1994.

Tuttle, Robert G. Jr. *John Wesley: His Life and Theology*. Grand Rapids, MI: Zondervan, 1978.

United Methodist Church. "An Apology for Support of Eugenics," no pages. http://calms.umc.org/2008/Text.aspx?mode=Petition&Number=1175.

United Nations. "Universal Declaration of Human Rights." December 10, 1948, no pages. http://www.un.org/en/universal-declaration-human-rights/.

Vacek, Heather H. *Madness: American Protestant Responses to Mental Illness*. Waco, TX: Baylor University Press, 2015.

Van Dyke, Joseph S. *Theism and Evolution: An Examination of Modern Speculative Theories as Related to Theistic Conceptions of the Universe*. New York: A.C. Armstrong & Son, 1886.

Vieler, Eric H. *The Ideological Roots of German National Socialism*. New York: Peter Lang, 1999.

Volf, Miroslav. "Living with the 'Other.'" In *Journal of Ecumenical Studies* 39 (2002) 8–25.

Wallis, Brian. "Black Bodies, White Science: Louis Aggasiz's Slave Daguerrotypes." In *American Art* 9 2 (1995) 39–62.

Wannenwetsch, Bernd. "'My Strength Is Made Perfect in Weakness': Bonhoeffer and the War over Disabled Life." In *Disability in the Christian Tradition: A Reader*, edited by Brian Brock and John Swinton, 353–90. Grand Rapids, MI: Eerdmans, 2012.

Ward, Lester Frank. *Neo-Darwinism and neo-Lamarckism*. Washington, DC: Gedney & Roberts, 1891.

Weikart, Richard. *Hitler's Ethic*. New York: Palgrave Macmillan, 2008.

Weismann, August. *The Germ Plasm: A Theory of Heredity*. New York: Charles Scribners Sons, 1893.

Whitney, Daniel W., and Kenneth N. Rosenbaum. "Recovery of Damages for Wrongful Birth." In *Journal of Legal Medicine* 32 (2011) 167–204.

Willard, Frances. "A White Life for Two, 1890." In *"Do Everything" Reform: The Oratory of Frances E. Willard*, edited by Richard W. Leeman, 159–72. Great American Orators 15. New York: Greenwood, 1992.

Wentz, Abdel Ross. *A Basic History of Lutheranism in America*. Philadelphia: Fortress, 1964.

Witkowski, J.A., and J. R. Inglis. *Davenport's Dream: 21st Century Reflections on Heredity and Eugenics*. Cold Spring Harbor, NY: Cold Spring Harbor Laboratory, 2008.

Wright, George Frederick. "The Debt of the Church to Asa Gray." In *Bibliotheca Sacra* 45 (1888) 523–30.

———. "The Passing of Evolution." In *The Fundamentals: The Famous Sourcebook of Foundational Biblical Truths*, edited by R.A. Torrey, 613–25. 12 vols. 1910–1915. Reprint, Grand Rapids, MI: Kregel, 1990.

———. "The Proper Attitude of Religious Teachers Toward Scientific Experts." In *New Englander* 37 (1878) 776–89.

———. *Studies in Science and Religion*. Andover, MA: Warren F. Draper, 1882.

Zenderland, Leila. "Biblical Biology: American Protestant Social Reformers and the Early Eugenics Movement." In *Science in Context* 11 (1998) 511–25.

———. *Measuring Minds: Henry Herbert Goddard and the Origins of American Intelligence Testing.* New York: Cambridge University Press, 1998.

Zimmerman, Jonathan. *Distilling Democracy: Alcohol Education in America's Public Schools.* Lawrence, KS: University of Kansas Press, 1999.

———. "'When the Doctors Disagree': Scientific Temperance and Scientific Authority, 1891–1906." In *Journal of the History of Medicine and Allied Sciences* 48 2 (1993) 171–97.

Zirkle, Conway. "The Early History of the Idea of the Inheritance of Acquired Characters and of Pangenesis." In *Transactions of the American Philosophical Society* 35 (1946) 91–151.

Made in the USA
Middletown, DE
04 March 2022

62135839R00124